# Water Conservation - Inevitable Strategy

*Edited by Murat Eyvaz, Ahmed Albahnasawi, Ercan Gürbulak and Ebubekir Yüksel*

Published in London, United Kingdom

IntechOpen

*Supporting open minds since 2005*

Water Conservation - Inevitable Strategy
http://dx.doi.org/10.5772/intechopen.95669
Edited by Murat Eyvaz, Ahmed Albahnasawi, Ercan Gürbulak and Ebubekir Yüksel

Contributors
Farid Alizad Oghyanous, Denis Nseka, Hosea Opedes, Frank Mugagga, Patience Ayesiga, Hannington Wasswa, Henry Semakula, Daneil Ochieng, Liudmila V. Vladimirovna Kireycheva, Valeriy M. Yashin, Ekaterina A. Lentyaeva, Aleksey D. Timoshkin, Harro Stolpe, Nguyen Ngoc Ha, Christian Jolk, Shefali S. Patel, Susmita Sahoo, Wided Mattoussi, Foued Mattoussi, Mohamed Salah Matoussi



First published in London, United Kingdom, 2022 by IntechOpen
IntechOpen is the global imprint of INTECHOPEN LIMITED, registered in England and Wales, registration number: 11086078, 5 Princes Gate Court, London, SW7 2QJ, United Kingdom
Printed in Croatia

British Library Cataloguing-in-Publication Data
A catalogue record for this book is available from the British Library

Additional hard and PDF copies can be obtained from orders@intechopen.com

Water Conservation - Inevitable Strategy
Edited by Murat Eyvaz, Ahmed Albahnasawi, Ercan Gürbulak and Ebubekir Yüksel
p. cm.
Print ISBN 978-1-80355-036-7
Online ISBN 978-1-80355-037-4
eBook (PDF) ISBN 978-1-80355-038-1

For EU product safety concerns:
IN TECH d.o.o., Prolaz Marije Krucifikse Kozulić 3, 51000 Rijeka, Croatia, info@intechopen.com or visit our website at intechopen.com.

# Meet the editors

Dr. Murat Eyvaz is an associate professor in the Environmental Engineering Department, Gebze Technical University, Turkey. His research interests include applications in water and wastewater treatment facilities, electrochemical treatment processes, filtration systems at the lab and pilot-scale, membrane processes (forward osmosis, reverse osmosis, membrane bioreactors), membrane manufacturing methods (polymeric membranes, nanofiber membranes, electrospinning), spectrophotometric analyses (UV, atomic absorption spectrophotometry), chromatographic analyses (gas chromatography, high-pressure liquid chromatography). He has co-authored many journal articles and conference papers and has taken part in many national projects. He serves as an editor and reviewer for many indexed journals. Dr. Eyvaz has four patents on wastewater treatment systems.

Dr. Ahmed Albahnasawi is a post-doctorate fellow in the Environmental Engineering Department, Gebze Technical University, Turkey. His graduate work focused on the investigation of the treatability of the sequential anoxic-aerobic batch reactors followed by ceramic membrane for textile wastewater treatment. Based on his Ph.D. research, Dr. Albahnasawi published three journal articles and participated in three international conferences. His research interests include the application and design of a microbial fuel cell integrated with Fenton oxidation for industrial wastewater treatment/solid waste management and monitoring of organic micropollutants by both chromatographic and spectrophotometric analyses.

Dr. Ercan Gürbulak is a research associate in the Environmental Engineering Department, Gebze Technical University, Turkey. He received his bachelor's degree in Environmental Engineering from Marmara University, Turkey, in 2005. He completed his MSc and Ph.D. at Gebze Technical University in 2008 and 2019, respectively. His research interests include the application and design of hydrothermal processes for industrial wastewater treatment/solid waste management and monitoring of organic micropollutants by both chromatographic and spectrophotometric analyses.

Prof. Ebubekir Yüksel is a faculty member of the Environmental Engineering Department, Gebze Technical University, Turkey. His research interests include applications in water and wastewater treatment facilities, electrochemical treatment processes, filtration systems at the lab and pilot-scale, watershed management, flood control, deep-sea discharges, membrane processes, spectrophotometric analyses, chromatographic analyses, and geographic information systems. He has co-authored numerous journal articles and conference papers and has taken part in many national projects. He has produced more than thirty peer-reviewed publications in indexed journals. He has one patent on pump/turbine design and four patents on wastewater treatment systems.

# Contents

# Preface

Water is essential for all living things. It covers 70% of our world and is also an important part of our bodies. However, only about 0.3% of the water resources on Earth is usable and potable. The use of this limited resource varies depending on demographic, economic, technological, and climatic characteristics. The consumption of water for domestic, industrial, and agricultural purposes has increased by 15% in the last two decades, and today one out of every three people is faced with the danger of water shortage. Since the population and water resources worldwide do not show a balanced distribution, nearly a hundred countries hosting almost half of the world's population suffer from water shortages.

Today's rapid population growth, water-consuming industries brought by technological developments, and epidemics spreading much faster due to globalization have greatly increased the need for water. In addition, global climate change, evaporation, and deterioration of rainfall balances also lead to negative consequences such as floods and droughts. It has been determined that surface and groundwater resources cannot meet the increasing demand for potable and thus used water and wastewater need to be recycled. However, issues such as protection of water resources, balanced water consumption, treatment of wastewater to a certain level, and realization of water recovery have shown that versatile water management should be introduced. To conserve water, both management and public strategies have been developed, alternatives have been revealed to reduce water consumption in industrial/commercial applications, and methods such as smart irrigation systems have been proposed in agriculture. Local authorities have focused on infrastructure operations to prevent water losses, and flow measurements have begun to be followed more closely. The use of greywater for partial recycling of water for household purposes and rainwater harvesting systems have started to be encouraged.

This edited volume consists of six chapters. Chapter 1 presents the general method of planning support in the areas of water and land use. Chapter 2 evaluates diffuse runoff from the drainage basin of a small river with agricultural land in the drainage basin and discusses the measures taken to protect the river waters from pollution. Chapter 3 examines the relationship between spatiotemporal land use/cover change and watershed health by using remotely sensed data. Chapter 4 identifies the main pollutants in different seasons and estuarine reaches by multivariate statistical methods, which are expected to help managers to understand the water body system along the estuary. Chapter 5 sheds light on the design of various incentive schemes to face groundwater over-exploitation by farmers who can over-pump water typically by manipulating their water meters in an asymmetric information context. Finally, Chapter 6 reviews the applications in which nanoparticles are used in water/wastewater treatment that play an indirect role in water conservation.

The editor would like to thank the authors for their contributions.

**Murat Eyvaz**
Associate Professor,
Department of Environmental Engineering,
Gebze Technical University,
Kocaeli, Turkey

**Ahmed Albahnasawi, Ercan Gürbulak and Ebubekir Yüksel**
Gebze Technical University,
Kocaeli, Turkey

Chapter 1

# Regional Water and Land Use Planning: Systematic Planning Support

*Harro Stolpe, Nguyen Ngoc Ha and Christian Jolk*

## Abstract

Sustainable water and land use planning is an important component of regional planning. Regional planning, as a multi-sectoral concept, serves as a framework for sub-regional and local planning. Water and land use planning needs as comprehensive a system understanding as possible, of water and land uses. This forms the basis for managing water resources quantity and quality, in coordination with the existing land use. The developed systematic GIS-based planning approaches, for a sustainable water allocation and water quality conservation, as well as the resulting recommendations for actions, supports water management on a regional and local planning level. The developed concepts are illustrated by up-to-date project results of the R&D research projects "ViWaT-Planning" Stolpe et al., in the Mekong Delta in Vietnam and "iWaGSS" Jolk et al., in the Olifants River Basin in South Africa. Both R&D projects are funded by the German Federal Ministry of Education and Research (BMBF) and are coordinated directly with existing water management institutions and their measures in both countries. The focus of this publication will be on the general methodology of planning support, in the fields of water and land use.

**Keywords:** regional water and land use planning, Mekong Delta, Olifants River

## 1. Introduction

Several research projects developed by the Institute of Environmental Engineering + Ecology (EE+E), at the Ruhr-University Bochum, assess water and land use at the regional level to support spatial planning. These research projects are located, mainly, in Vietnam and in South Africa, under different spatial and climatic conditions. **Table 1** characterizes the different investigation areas [1–5].

In each R&D project, the central core of the work was to analyses the current water and land use system, in a transparent and systematic way. This analysis aids in designing suitable management measures for water resources, with regards to water quantity and water quality.

Based on the investigated current state of water management in the different water systems, problems and problem areas are identified and needs for action are described and prioritized, on a regional scale. The results of the aforementioned R&D projects, is a systematic planning support system, for identifying problems and conflicts, in water and land use management.

Systematic water and land use planning is understood as a continuous process that must be able to respond to changing natural conditions and land use

| Location | Hydrological system | Climate | Water system | Objectives | R&D project |
|---|---|---|---|---|---|
| Vietnam | | | | | |
| Red River | Red River Delta | Temperate, dry winter, hot summer | Polder management | Water quantity, water quality | IWRM |
| Mekong | Mekong Delta | Tropical, savannah | Open canal system, weirs, sluice gates | Water quantity, water quality | IWRM; VIWaT Planning |
| Vu Gia Thu Bon | Vu Gia Thu Bon lowland | Tropical, monsoon | River, open canal system | Water quantity, flood, salinity | LUCCI, WaLaMa |
| Dong Nai | Middle Vietnam highlands | Tropical, monsoon | Dam management | Water quantity, water quality | IWRM |
| Dong Van | Northern Vietnam highlands | Temperate | Karst | Water quantity, water quality | Kawatech |
| South Africa | | | | | |
| Olifants | High Plateau | Arid | Dam management | Water deficit management, water quality | MOSA, iWaGSS |

**Table 1.**
*Regional water and land use projects in Vietnam and South Africa, performed by the EE+E.*

innovations. Updates are, therefore, required from time to time. Such updates can be done, effectively, within a geodata management system. Such a system has to involve the planning levels relevant to water and land use issues:

- The national level provides a framework of laws and technical standards for water and land use to be managed at the regional or local level.

- The regional level is used to name specific water and land use management measures, on the basis of a regional framework for planning and detailed inquiries, in order to remedy discovered problems and to make decisions on specific locations. The regional level has a special significance due to spatial and thematic interactions.

- The local level is used to carry out object planning at the previously identified locations.

This top-down process enables the higher levels to influence the lower levels. Conversely, it should be accompanied by bottom-up processes, for the lower levels to be considered at the higher levels. Only in this way can a continuous planning process be adapted to the constantly evolving innovations.

The basis for such a planning system approach is a suitable collection, harmonization and evaluation of the relevant water and land use related information, as well as a generally comprehensible presentation and visualization of the various planning issues, to support decision making. This process is to ensure the scientific soundness and application of advanced technologies, the connectivity, the forecasting ability, the feasibility, and the effective and thrifty use of the land and water resources; the objectivity, publicity, transparency and conservation (e.g. in Vietnam: Clause 5, Article 4, Law on Planning No. 21/2017/QH14).

The methodologies developed in the different R&D projects (see **Table 1**), used modern techniques of data acquisition, harmonization and processing, as well as methods to evaluate and visualize the results with the help of Geographic Information Systems (GIS).

This publication summarizes experiences gained from several research projects, funded by the German Federal Ministry of Education and Research (BMBF), related to data management and spatial decision support.

All projects were carried out in close cooperation with a high number of responsible institutions in Vietnam (e.g. MOST Ministry of Science and Technology, MONRE Ministry of Natural Resources and Environment, NAWAPI National Center of Water Resources Planning and Investigation as well as local responsible authorities) and in South Africa (e.g. DWA Department for Water Affairs, DWS Department of Water and Sanitation, The WRC Water Research Commission, Kruger National Park and SAEON South African Environmental Observation Network, etc.).

## 2. Methodological approach

The central objectives of the R&D projects, listed in **Table 1**, was to develop a systematic and updatable GIS-based planning support approach, to describe and analyses the current state, future states and the respective problems, on water and land use management.

A prerequisite to this work, is to acquire and develop a structured collection of the relevant data on water and land use. In both countries Vietnam and South Africa, a large amount of relevant data exists, but it is not always systematically collected, stored and evaluated.

The most important step in the methodological approach is to store the scattered data sets of varying topicality (administration, topography, population, water demand, water resources, water quality, sensitivity of water resources, etc.) in one centralized location, for example, a geodatabase. Additional steps include the harmonization of the data sets as well as to check the data sets on their plausibility.

**Table 2** shows some of the varying constraints to data collection in Vietnam and South Africa, on which the sometimes-time-consuming initial phase of regional water and land use projects, depends.

Based on the collected data and information, including the aforementioned harmonization and plausibility checks, a water and land use planning information system, as well as GIS-based water and land use planning maps for decision makers, were developed.

Based on the systematically collected, stored and evaluated database different methods to analyse water quantity as well as water quality, were developed (see **Figure 1**).

For water quantity aspects, methods were developed to compare water demands/water uses with water resources:

| Data | Vietnam | South Africa |
|---|---|---|
| Data availability | No Open Data Act | Open Data Act |
| | Distributed over institutions | Distributed over institutions |
| Data formats | Several different data formats | Shape Data, Google Earth |
| Data consistency | Low to medium quality | High to medium quality |
| Metadata | Non-existent | Sparse |
| Costs | Usually include an acquisition fee | Generally available at no cost |

**Table 2.**
*Different data handing in Vietnam and South Africa.*

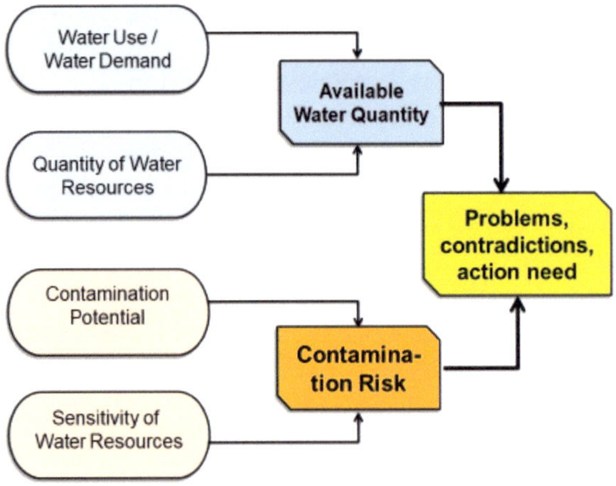


**Figure 1.**
*Structure of the water and land use planning system and planning maps.*

- Water demand/water use: related to different land uses (agriculture, aquaculture, settlements, mining, industry, etc.)

- Water resources: surface water, groundwater

- Result of comparison: available water quantities, water deficits, water surpluses, water allocation

For water quality aspects, methods were developed to combine contamination potentials of land uses (agriculture, settlements, mining, industry, etc.), with the sensitivities of the water resources (surface water, groundwater, ecology).

- Contamination potentials: different land uses are responsible for various contamination potentials for surface water and groundwater, depending on chemical substances used (e.g. fertilizers, pesticides, etc.) and their application methods, as well as agricultural practices, themselves.

- Sensitivity of water resources: sensitivity of water resources are varied, according to the boundary conditions for surface water (flow velocity, pollutant degradation, extraction practices and their proximity to sensitive resources, etc.), for groundwater (filtering effect of the covering soil layers, extraction practices and their proximity to sensitive resources, etc.) and the ecological status of the resources.

- Result of comparison: the result of the combination is the designation of different contamination risks (e.g. high contamination potential + high sensitivity = high risk or low contamination potential + low sensitivity = low risk, etc.).

Based on the systematic approach outlined in **Figure 1**, a number a different water resources planning maps, have been developed. They represent water quantity as well as water quality issues. All the thematic maps were visualized in water planning maps (in some cases bundled in a Water Management Atlas or in a Web-GIS system [5]).

An important task was to design the work in such a way that it does not depict conditions that are valid in the short term, but rather derives planning-relevant statements that are valid in the longer term.

This publication discusses examples for the determination of the usable water quantity in the Mekong Delta, in Vietnam and for the determination of contamination risks in the Olifants river basin, in South Africa.

### 2.1 Available water quantity (example of Vietnam)

In accordance with the aforementioned planning levels, at the national level in Vietnam, the Planning Law No. 21/2017/QH14 [6], 2017, the Resolution No. 120/NQ-CP, 2017 [7] and the Decree No. 37/2019/ND-CP [8] are of basic importance in the context of water and land use planning. At the regional level, Decree No. 37/2019/ND-CP is of importance.

Based on these legal preconditions, the starting point of any planning-oriented water and land use management consideration, is to carry out an adequate system analysis. The system analysis should present the essential qualitative and quantitative interdependencies, as a basis for the planning assessments.

The project area, situated within the Mekong Delta (see **Figure 2**), is located in the South of Vietnam. It includes seven provinces, south-west of the Hau River, with an area of 22,000 km$^2$ and contains approximately nine million inhabitants.

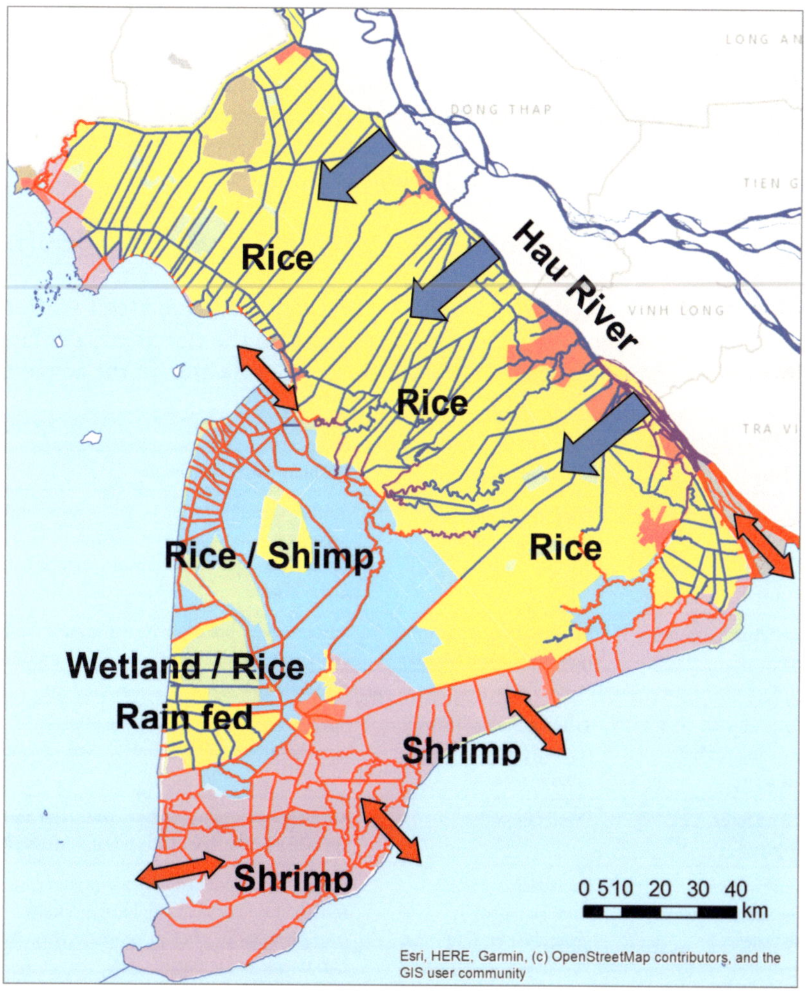


**Figure 2.**
*Mekong Delta water and land use system in the dry season (Arrows: blue = fresh water from Hau River, red = saline water exchange/intake. Water bodies: blue = fresh, red = saline, green = alternating fresh/saline).*

The Mekong Delta is a tropical delta area with terrain elevations mainly between 0 and 50 cm A.S.L. The Delta is a complex system of rivers, canals, rice paddies, aquacultures and wetland ecosystems, as well as infrastructure for water supply and irrigation.

The prominent feature of the Mekong delta is the existing dense water network of mostly human-made canals with various functions, including: ship transportation, irrigation, waste water discharge, water storage, etc. Characteristic for the Mekong Delta is the existing effective regional freshwater and saltwater management, controlled by weirs and sluice gates, with different water management functions, such as fresh water intake, fresh water and saline water separation, saline water intake and saline water discharge. These functions enable or constrain different land uses (rice and fruit production in freshwater areas, shrimp aquaculture in saltwater areas).

The driving forces for this water management are the tides and the changing water flows during the rainy and dry seasons. The tides and seasonal water levels allow for a recurrent exchange of fresh and salt water in the Mekong Delta.

The agricultural areas, such as rice and fruit production, are fed by freshwater from the Hau River. The areas with saline water from the sea are characterized by saltwater fish and shrimp farming.

Current trends show a significant decline in rice areas and an increase in aquaculture and settlement areas. Natural areas are only present in remaining areas. The land use planning under the first Mekong Delta Integrated Regional Plan (MDIRP) reinforces this trend as the influence of salt increases due to sea level rise, land subsidence and the decline of inflows into rivers.

### *2.1.1 Water demand, water resources in dry season*

In the following table, the focus is on the dry season, as this is the critical time for ensuring water supply.

Within the study area, there are different water demands related to different land uses. There is a demand for fresh surface water in the rice areas, a demand for saline seawater, as well as a demand for fresh groundwater, in the aquaculture

| Land use | Water resources | Problems/measure need |
|---|---|---|
| Paddy rice—fresh water | Fresh surface water from Hau River | Surface water pollution by waste water and agrochemicals; water shortages at the Hau River mouth during saline intrusions; lack of fresh water in dry season |
| Aquaculture—saline water | Saline sea water, fresh rain water, groundwater | Surface water pollution by waste water and aqua chemicals; lack of fresh water; local groundwater over-exploitation |
| Rice/shrimp—alternating fresh or saline water | Saline sea water in dry season, fresh rain water in rainy season | Surface water pollution by waste water, agrochemicals, aqua chemicals; soil salinization |
| Wetland forest/rice—fresh water | Fresh rain water | Surface water pollution by waste water and agrochemicals; lack of fresh water in dry season |
| Urban and rural settlements, industry—fresh water | Surface water (e.g. from Hau River), groundwater, rain water | Surface water pollution by waste water; lack of fresh water in dry season; local groundwater over exploitation, lack of space for rain water harvesting and storage |

**Table 3.**
*Main water and land uses in the Mekong Delta according to the map in* **Figure 2** *water demand, water resources, problems and measure need.*

| Land use | Area (ha) | Area (%) | Water demand (million m$^3$/a) | Water demand (%) |
|---|---|---|---|---|
| Triple rice | 135,993 | 52 | 1.353 | 60.9 |
| Double rice | 83,148 | 32 | 0.794 | 35.7 |
| Rivers, canals, water surfaces | 13,742 | 5 | — | — |
| Residential land in rural area | 8158 | 3 | 0.014 | 0.6 |
| Residential land in urban area | 5461 | 2 | 0.013 | 0.6 |
| Others | 14,788 | 6 | 0.047 | 2.1 |
| **Total rice, residential** | **246,502** | **94** | **2.174** | **97.9** |
| Total all land uses | 261,290 | 100 | 2.221 | 100 |

**Table 4.**
*An example of land uses and water demands from so-called subregion III, a freshwater area in the northwest of the study area (see* ***Figure 2****) near the border to Cambodia.*

areas and alternating water demand (saline seawater and rainwater) in the rice-shrimp areas. Additionally, there is a demand for rainwater in the rainfed wetland/rice areas. **Table 3** provides the basic information on water resources and the associated water demand for the main land use types (see **Figure 2**), in the Mekong Delta.

In determining the water demand, the following water use sectors are considered: agriculture, aquaculture, domestic water, drinking water, etc. A method for estimating total water demand, according to the current land use classifications by the Vietnamese Ministry of Natural Resources and Environment (MONRE), was developed and applied.

In the freshwater areas along the Hau River in the north-east of the study area, the main user of large quantities of freshwater, in the dry season, is rice cultivation (see **Table 4**). The demand for water can be supplemented by water storage in small canals, from the rainy season, as well as inflows from the Hau River. The water demand for residential land uses, in this area, is covered by surface water from the Hau River or groundwater.

In the saline areas in the southwest, where there is no relevant demand for fresh water for irrigation, the main sources of freshwater come from groundwater extraction, rain water harvesting as well as storage (e.g., for domestic purposes, food processing, etc.).

### *2.1.2 Problems, contradictions, need for action*

The following problems with different land use and their associated water demands, become apparent:

- Quality problems and limitations of the usability of surface water resources occur due to pollutant inputs (sewage, agrochemicals, aqua chemicals, etc.)

- Quantity problems occur due to episodic saltwater intrusions, in the Hau River mouth. During such periods, water abstraction for irrigation purposes is not possible.

- There is a freshwater demand in the saline areas that cannot be met by rainwater, as there are too few areas and facilities to collect the rainwater in rainy season. This leads to overuse of groundwater in these areas.

- The demand for groundwater in the saline areas, leads to local over-exploitation of groundwater resources, which leads to land subsidence.

- The current problems are exacerbated by a greater diversity of water users with different water needs, and thus the quantitative increase in water demand.

- There is also a decline in natural freshwater ecosystems and loss of ecological functions within the rivers and canal systems.

Water demand, water use and water resources, as well as the resulting problems and contradictions, are depicted in corresponding planning maps and underpinned by results from mathematical models. Proposals for the avoidance and improvement of water and land use are derived from this.

## 2.2 Contamination risks (example of South Africa)

South Africa is facing major challenges in the water sector. The uneven distribution within river networks and insufficient precipitation leads to water-supply shortages, especially in dry season. Additionally, water infrastructure and the management of water supply and waste water treatment, is in deficit. The rapid industrial growth, the progressing urbanization and the industrially organized agriculture, lead to increasing water demand and water quality problems [2].

The investigated Olifants River Basin is located in the provinces Limpopo, Gauteng and Mpumalanga, in the northeast of South Africa (see **Figure 3**). It has a total area of 54.626 km$^2$.

In contrast to the Mekong Delta, the Olifants River Basin has a semi-arid climate and, therefore, stressed water resources. Economically, it is characterized by intensive mining activities, irrigated and rainfed agriculture and tourism (Kruger National Park). Surface water, stored in dams and reservoirs, supplements most of the water demand. The largest dams are: Loskop Dam and De Hoop Dam (see **Figure 3**).

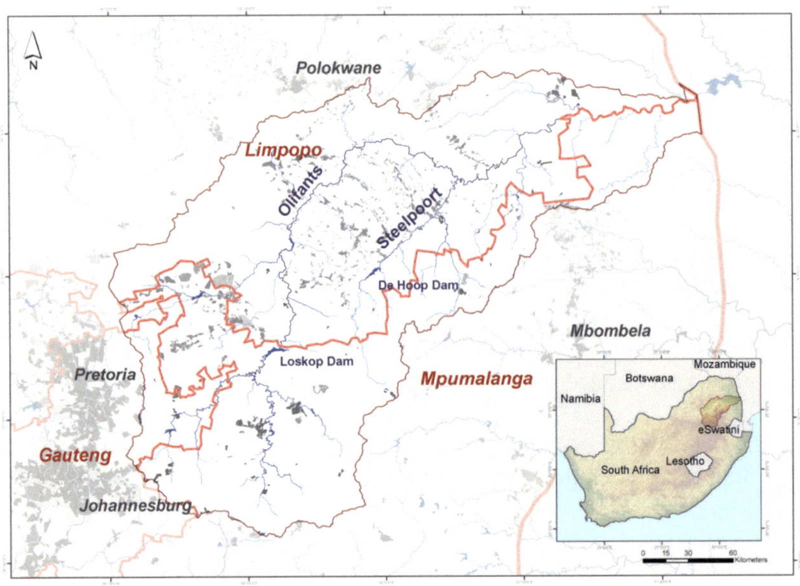


**Figure 3.**
*Olifants River Basin.*

*DOI: http://dx.doi.org/10.5772/intechopen.102713*

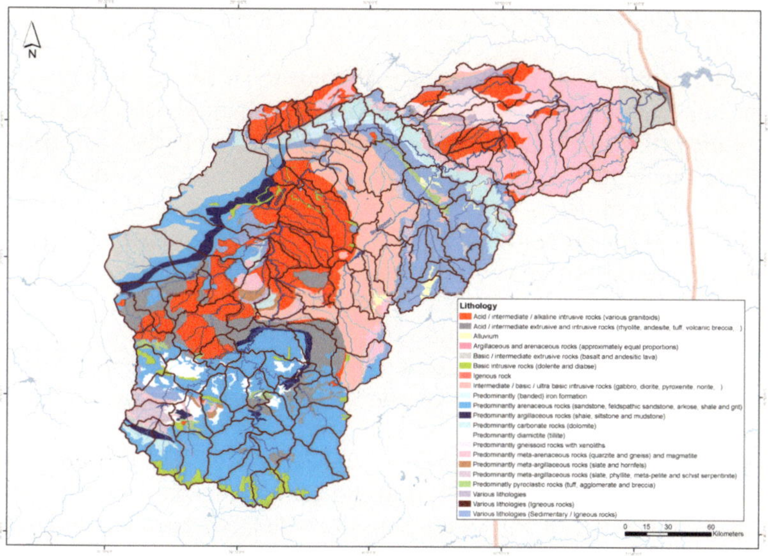


**Figure 4.**
*Lithology [9–12].*

The Olifants River Basin is characterized by two main river systems: the Olifants and Steelpoort Rivers (see **Figure 3**). Groundwater use is limited to only one exploitable aquifer, formed by basalts. The rest of the river basin consists primarily of sedimentary rocks (e.g. sandstone) and intrusive igneous rocks (e.g. granite) (see **Figure 4**).

Water quality issues are a limiting factor for the socio-economic and environmental development in South Africa. Many studies in South Africa focus on the negative side-effects of dynamic population and economic growth, industrially organized agricultural practices and contamination potential, originating from the mining sector. In addition to these topics, inefficient water infrastructure is also one of the main problems [2].

The Water Quality Report of the Department of Water Affairs [13] and the Planning Level Review of Water Quality in South Africa [14], provides an assessment of the existing conditions of water quality in the Olifants River Basin.

According to Van Veelen [13], the analysis results of the water quality in the Olifants River Basin highlight the following: salinity related impacts due to mining, power generation and industries; unacceptable EC concentrations due to irrigation return flows; acid mine drainage; mesotrophic to eutrophic dams; unacceptable phosphate concentrations in rivers; unacceptably high levels of heavy metal concentration in parts of the catchment; pesticides and herbicides in rivers due to agricultural activities.

This wide range of water quality-related problems, needs a holistic and efficient research approach. The method developed for assessing contamination risks, includes consideration of the following contamination pathways:

- Infiltration of pollutants into groundwater
- Erosive surface runoff of pollutants into surface water bodies
- Direct discharge of pollutants into surface waters

In this publication, only the sensitivity of the groundwater resources and the contamination potentials of agriculture, are determined and aggregated, to contamination risks on a sub-basin level.

### *2.2.1 Sensitivity of groundwater resources*

**Figure 4** illustrates the lithology of the Olifants River Basin. The runout evaluation of the uppermost groundwater aquifer was carried out, based on the lithological classes and is described via their hydraulic conductivity in the hydrogeological map series of the Republic of South Africa [9–12]. Overlying strata above the aquifers were not considered, as these cannot be safely assessed at the chosen scale of 1:1,300,000. As the protective effect of overlying strata is ignored, the classification of resource sensitivity lies within secure margins. The consideration of overlying strata is to be included in subsequent planning levels.

The resource sensitivity of groundwater is established for the uppermost groundwater aquifer, also taking into consideration groundwater use. Groundwater resource sensitivity is classified, based on runout and groundwater use as per **Table 5**.

**Figure 5** depicts the groundwater resource sensitivity in the Olifants River Basin. Areas with groundwater extraction and use are characterized by very high resource sensitivity. They are shown, using a hatched pattern. This characterization is based on existing risks through contaminant inflow, due to informal and illegal borehole construction or borehole use. A further justification for this characterization is the special need for protection of the directly used groundwater resource. In accordance with the Technical Rules for the Protection of

| Very high groundwater sensitivity | High groundwater sensitivity | Medium groundwater sensitivity | Low groundwater sensitivity |
|---|---|---|---|
| Increased groundwater use area | Solid rock—high yield (e.g., basalt); floodplains with low depth to water table | Solid rock—medium yield (e.g., alternating layers of sandstone, shale); loose material—medium yield (e.g., sand) | Solid rock—low yield (e.g., granite) |

**Table 5.**
*Evaluation of groundwater resource sensitivity in the Olifants River Basin [9–11].*

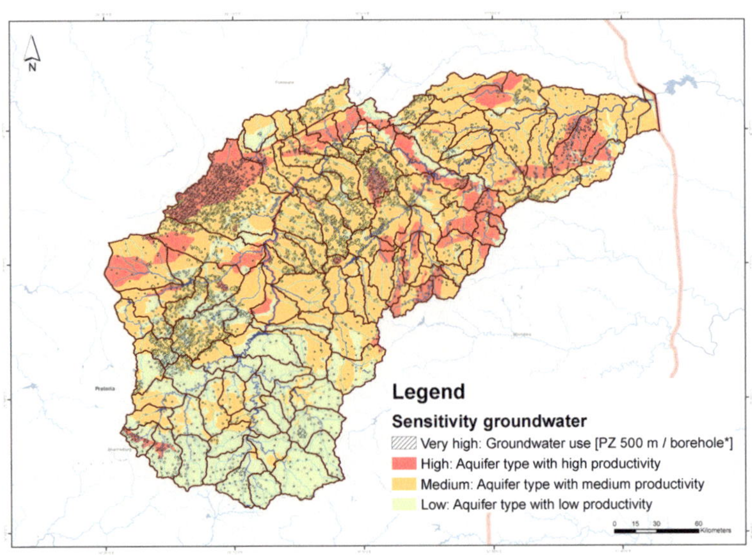


**Figure 5.**
*Sensitivity groundwater.*

Groundwater (Protected Areas for Groundwater, W101 [15, 16]), which were drawn up in Germany, by the DVGW, together with LAWA, a protection zone of 500 m around each borehole was defined, in this study, for the project area.

In **Figure 5**, areas with the classification "high" resource sensitivity are depicted in red, those with "medium" resource sensitivity are depicted in orange and those with "low" resource sensitivity are depicted in green.

### *2.2.2 Contamination potentials of land uses*

Distinction is made between the following possible contamination potentials:

- Contamination potential from diffuse sources, through infiltration of agricultural contaminants into groundwater
- Contamination potential from diffuse sources, through infiltration of waste water from settlements into groundwater
- Contamination potential from point sources, through infiltration of contaminants into groundwater (commercial, industrial, dumpsites, mines)

In the following sub-chapter, the contamination potential from diffuse sources, through infiltration of agricultural contaminants into groundwater, will be discussed.

For the different land use classes in the Olifants River Basin, contamination potential through infiltration of nutrients into groundwater, is assumed. **Figure 6** depicts land cover and its classification, based on the South African National Land-Cover Database [17]. The nutrient availability potential is differentiated, according to different land cover classes, analysed by Moolman [18], for the Olifants River Basin (see **Table 6**).

In a next step, the results were divided into three contamination potential classes. Areas with a "high" nutrient available potential are depicted in red, those with a "medium" nutrient availability potential are depicted in orange and those with "low" nutrient availability potential are depicted in green (see **Figure 7**).

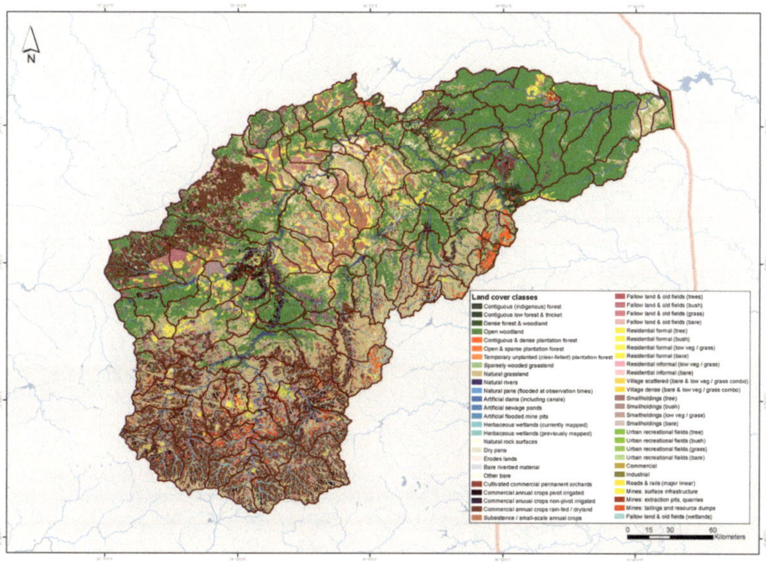


**Figure 6.**
*Land cover (72 classes) [17].*

| Land use classes | Contamination potential |
|---|---|
| Bare Rock and Soil (erosion: dongas/gullies) | Medium |
| Bare Rock and Soil (erosion: sheet) | Medium |
| Bare Rock and Soil (natural) | Low |
| Cultivated: permanent—commercial dryland | High |
| Cultivated: permanent—commercial irrigated | High |
| Cultivated: temporary—commercial dryland | High |
| Cultivated: temporary—commercial irrigated | High |
| Cultivated: temporary—semi-commercial/subsistence dryland | High |
| Cultivated: temporary—semi-commercial/subsistence irrigated | High |
| Degraded: forest and woodland | Medium |
| Degraded: thicket and bushland (etc.) | Medium |
| Degraded: unimproved (natural) grassland | Low |
| Forest (indigenous) | Low |
| Woodland | Low |
| Forest plantations (Acacia spp.) | Medium |
| Forest plantations (deforestation) | Medium |
| Forest plantations (Eucalyptus spp.) | Medium |
| Forest plantations (other, mixed spp.) | Medium |
| Forest plantations (Pine spp.) | Medium |
| Improved grassland | Medium |
| Thicket, Bushland, Bush Clumps, High Fynbos | Low |
| Unimproved (natural) grassland | Low |
| Wetlands | Low |

**Table 6.**
*Contamination potential land use classes [18].*

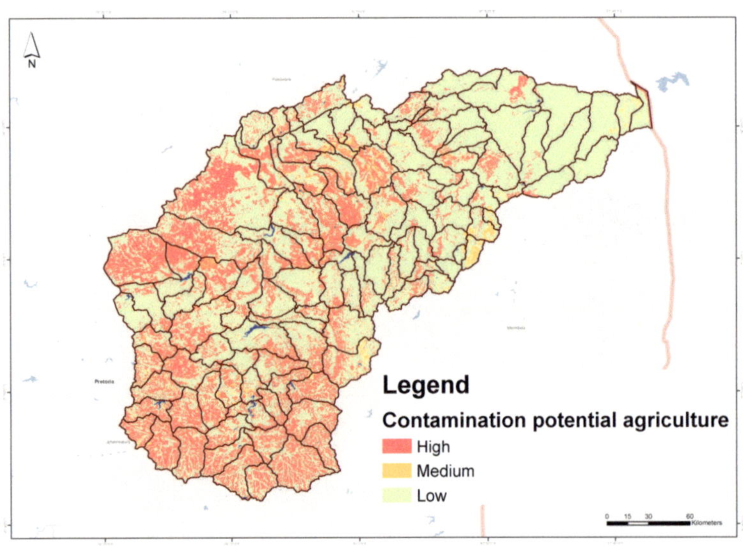


**Figure 7.**
*Contamination potential of diffuse sources, due to infiltration of agricultural contaminants into groundwater.*

| | | | | | |
|---|---|---|---|---|---|
| **Sensitivity groundwater** | Very high | No | Very high | Very high | Very high |
| | High | No | Medium | High | High |
| | Medium | No | Low | Medium | High |
| | Low | No | Low | Low | Medium |
| **Contamination risk agriculture** | | None | Low | Medium | High |
| | | **Contamination potential agriculture** | | | |

**Table 7.**
*Groundwater contamination risk through infiltration of agricultural contaminants.*

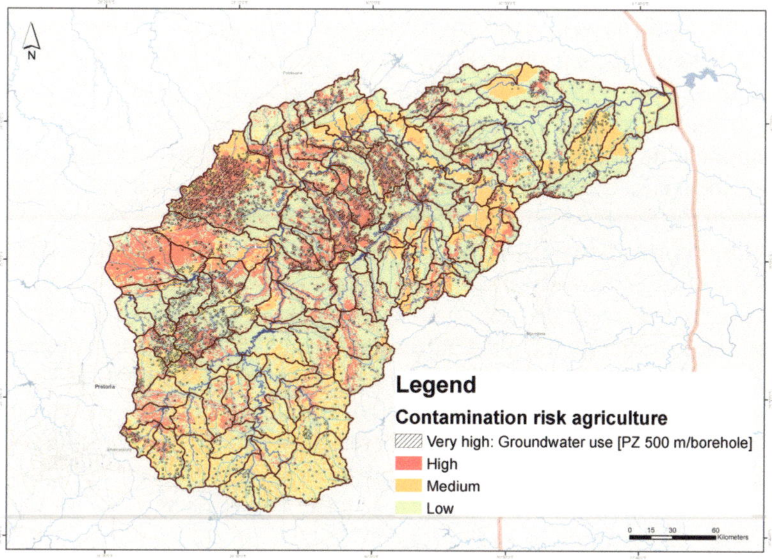


**Figure 8.**
*Groundwater contamination risk due to infiltration of agricultural contaminants into groundwater.*

### *2.2.3 Groundwater contamination risk*

The groundwater contamination risk classification, is the result of aggregating the aforementioned groundwater resource sensitivity and the corresponding contamination potential.

**Table 7** represents the aggregation method used to establish the contamination risk for groundwater resources, from agriculture. These contamination risks are classified as "very high", "high", "medium" and "low".

**Figure 8** depicts the groundwater contamination risk in the Olifants River Basin. The contamination risk of areas depicted in 44 red is "high", the contamination risk of areas depicted in orange is "medium" and areas depicted in green have a "low" contamination risk.

Overall, the results for the Olifants River Basin shows large areas with an increased contamination risk. Vast areas with an increased contamination risk are highly relevant for water supply (boreholes) management interventions. A reduction of the contamination risk, for example through water protection concepts, is necessary.

## 3. Conclusions and outlook

The discussed methodology was developed and applied, under different boundary conditions, in several river basins in Vietnam and South Africa.

Depending on the available information and data, the methods used to evaluate the water resources quantity and quality, were adapted to the different boundary conditions. The methods were also integrated into the existing spatial water and land use planning concepts.

This methodology is found to be suitable for the identification of problems and need for action, at the regional level and as an orientation for sub regional/sub-basin levels, as well as water management units, with their local investigations and measurements protocols.

The thematically-oriented collection of geodata, related to water and land use, will create a centralized geodatabase, in which all relevant information will be available, across varying institutions and disciplines.

The described methodology analyses and compares water demands and water resources, in detail, as well as sensitivities of water resources and contamination potentials. Results of the methodology, define existing and expected problems and needs for action, on a river basin and sub-basin level, with regard to the availability of water resources and pollution risks.

The application of the methodology will enable stakeholders to make decisions, on a scientific basis. The identification process enables decision makers to attend effectively to the issues with high priority ratings first.

The close cooperation with the Vietnamese and South African authorities, within the projects, ensures a holistic implementation of the different methods on a local level. The participation of the responsible water agencies, on national level, guarantees a sustainable adjustment and a nation-wide transferability of the methods.

The derived results, from the developed methods, are documented in several reports and documents [4, 19–21]. This publication has outlined the methodology developed. Further publications show concrete applications for planning support at the regional level in Vietnam and in South Africa. This refers to the main planning maps (see **Table 8**), related explanations and recommendations for measures (monitoring, use restrictions, sanitation, etc.).

In addition, the GIS-based methods were transferred to web-based GIS systems. The software tools complement the sustainable use of the methodology. They

| | **Regional planning support—Planning Maps** | **MD/ORB** |
|---|---|---|
| Water quantity | Fresh SW demand according land uses | MD |
| | Available fresh SW resources considering water quality | MD |
| | Deficits, surpluses of fresh SW | MD |
| | Fresh GW demand according land uses and GW extractions | MD |
| | Available fresh GW resources | MD |
| | Deficits, surpluses of GW | MD |
| Water quality | Contamination Potentials for fresh SW | MD |
| | Sensitivity of fresh SW resources | MD |
| | Contamination risks for fresh SW, monitoring results of SW | MD |
| | Contamination potentials for fresh GW | MD, ORB |
| | Sensitivity of fresh GW resources | MD, IRB |
| | Contamination risks for GW, monitoring results of SW | MD, ORB |

*SW = surface water, GW = groundwater, MD = Mekong Delta, ORB = Olifants River Basin.*

**Table 8.**
*Outcome: main GIS-based updatable planning maps.*

facilitate easy-to-use, standardized and quality-assured techniques for updating data and integrating new data. In a user-friendly way, the software tools also make the planning maps available, countrywide, through a web browser.

The main users of the methods and systems should be, for example, institutions for coordination of regional planning or river basin organizations. The web-based systems developed provide an important comprehensive information basis. Based on this, decision-makers can make decisions in a transparent and comprehensible manner.

## Author details

Harro Stolpe[1]*, Nguyen Ngoc Ha[2] and Christian Jolk[3]

1 EE+E—Environmental Engineering and Environment, Ruhr-University of Bochum, Bochum, Germany

2 NAWAPI—National Center of Water Resources Planning and Investigation, Hanoi, Vietnam

3 Technische Hochschule Ostwestfalen-Lippe, Höxter, Germany

*Address all correspondence to: harro.stolpe@rub.de

## References

[1] Greassidis S, Borgmann A, Führer N, Jaschinski S, Jolk C, Stolpe H, et al. Überschlägige Wasserbilanz zur Planungs- und Entscheidungsunterstützung auf Einzugsgebietsebene für das Integrierte Wasserressourcen-Management in Vietnam, Hydrologie und Wasserbewirtschaftung, 55. Jahrgang, Heft 2, April 2011. Koblenz, Germany: Water Resource Management, Hydrology and Water Management; 2011

[2] Jolk C, Zindler J, Stolpe H, Zindler B, Wössner R, Abecker A. Planning and decision support tools for integrated water resource management on the river basin level in South Africa using the example of the Middle Olifants sub-basin. In: Prof. Dr.-Ing. Dr. rer. Pol. Dr. h. c. K.-U. Rudolph: MOSA integrated water resources management in the "Middle Olifants" river basin, South Africa – Phase II Summary Report. IEEM gGmbH: Schriftenr. Umwelttechnik u. Umweltmanagement. 2016;**34**(1):17-73

[3] Stolpe H et al. Regional Planning in the Mekong Delta – The System: Freshwater, Saline Water, and Land Uses Determines – Planning Questions – The R&D Project ViWaT-Mekong-Planning Develops Planning Support Tools. Vietnam: Hanoi; 2021. Available from: https://watersecurity.info/wp-content/uploads/2021/04/WSCC_BookofAbstracts_2021.pdf

[4] Stolpe H et al. Method Handbook for IWRM in Vietnam on River Basin Level. Vietnam: Hanoi; 2013

[5] Zindler B, Borgmann A, Greassidis S, Jaschinski S, Jolk C, Stolpe H. Planning and decision support tools for integrated water resources management (IWRM) on River Basin Level in the Southeast-Asian Region on the Example of Vietnam: Tools for water quantity and quality risk assessment. In: In: Luo, Yuzhou: Novel Approaches and Their Applications in Risk Assessment. Rijeka: InTech; 2012. pp. 37-58

[6] The National Assembly of the Socialist Republic of Vietnam. The Law on Planning, Law No.: 21/2017/QH14, November 24, 2017. Vietnam: Hanoi; 2017

[7] The Government of the Socialist Republic of Vietnam. Decree Elaboration of The Law on Planning, No. 37/2019/ND-CP dated May 05, 2019. Vietnam: Hanoi; 2019

[8] The Government of the Socialist Republic of Vietnam. Resolution On sustainable and climate-resilient development of the Mekong delta, No. 120/NQ-CP dated November 17, 2017. Vietnam: Hanoi; 2017

[9] Barnard HC, Baran E. Hydrogeological Map Series of the Republic of South Africa. Pretoria: Department of Water Affairs and Forestry; 1999

[10] Du Troit AJI, Du Troit WH, Jonck F. Hydrogeological Map Series of the Republic of South Africa. Pretoria: Department of Water Affairs and Forestry; 1999

[11] Du Troit WH, Jonck F, Mullin H. Hydrogeological Map Series of the Republic of South Africa. Pretoria: Department of Water Affairs and Forestry; 1998

[12] Du Troit AJI, Du Troit WH, Jonck F. Hydrogeological Map Series of the Republic of South Africa. Pretoria: Department of Water Affairs and Forestry; 2003

[13] Van Veelen M, Dhemba N. Development of a Reconciliation Strategy for the Olifants River Water

Supply System. Water Quality Report. P WMA 04/B50/00/8310/7. Pretoria: DWA; 2011

[14] DWA. Planning Level Review of Water Quality in South Africa. Pretoria: Water Quality Planning; 2011 (P RSA 000/00/14010)

[15] DVWG. Technische Regel—Arbeitsblatt DVGW W 102: Richtlinien für Trinkwasserschutzgebiete; Teil 2: Schutzgebiete für Talsperren. Bonn, Germany: DVGW; 2002

[16] DVWG. Technische Regel—Arbeitsblatt DVGW W 101 (A): Richtlinien für Trinkwasserschutzgebiete; Teil 1: Schutzgebiete für Grundwasser. Bonn, Germany: DVGW; 2020

[17] Geoterraimage. South African National Land Cover Dataset (2018). Pretoria: DEA; 2018

[18] Moolman J, Quibell G, Hohls B. A Qualitative (GIS based) Model of Nonpoint Sources Areas. Modelling Suspended Sediment in the Olifants River Catchment. Pretoria: Department of Water Affairs & Forestry. Institute for Water Quality Studies; 1999. Available from: http://www.dwaf.gov.za/iwqs/reports/slopes_olifants/sed_olif.htm [Last checked: 04.02.2014]

[19] Jolk C et al. GIS-Based Planning Tools for Ingegrated Water Resources Management in South Africa. E-Proceedings of the 36th IAHR World Congress. Netherlands: Den Haag; 2015

[20] Jolk C, Wiggett J, Stolpe H. Kontaminationsrisikobewertung auf Flusseinzugsgebietsebene am Beispiel des Olifants in Südafrika. In: KA Korrespondenz Wasserwirtschaft, KW 10/2020. Hennef, Germany: DWA; 2020

[21] Wiggett J, Jolk C. Developing Early Warning System Support for the Lower Olifants River Basin using a 1D-hydrodynamic model, WISA 2020 Online Conference, December 07-11. South Africa: Johannesburg; 2020

Chapter 2

# Diffuse Runoff from Agricultural Lands within a River Basin and Water Protection Measures

*Liudmila V. Kireicheva, Valery M. Yashin, Ekaterina A. Lentyaeva and Aleksey D. Timoshkin*

## Abstract

This paper is dedicated to the study of the pollutants coming from agricultural lands located within the catchment into the Yakhroma river, a third-order tributary of the Upper Volga. The area of the river catchment is 1437 $km^2$. It is located in the north-eastern part of the Moscow region, which geographically belongs to the Klinsko-Dmitrov ridge (the upper part of the basin) and the Upper Volga Lowland. The slopes and floodplain included in the reclaimed lands (more than 9 th ha) are lined with cities, rural settlements, numerous kitchen gardens, and agricultural lands. Water quality, river profile from the source to the mouth, and sources of pollution within the reclaimed lands of the Yakhroma floodplain were studied from 2004 to the present. A geospatial intelligence system (GIS) was developed for the catchment area. Land areas are allocated according to the conditions of surface runoff formation, taking into account soil types and slopes. The studies of the river water quality, tributaries, and drainage network in the reclaimed lands showed biogenic pollution caused by insufficiently treated wastewater discharged from cities and agricultural land, especially within the reclaimed massif. The calculations of the removal of nitrogen, phosphorus, and potassium from surface and drainage waters revealed that the main role in the pollution of both surface and drainage waters is played by nitrogen and potassium compounds, and to a lesser extent by phosphorus compounds. For nitrogen, removal from surface runoff was 27.36 t/year; for phosphorus it was 6.06 t/year; for potassium it was 242.28 t/year; with drainage runoff, the removal of nitrogen was 98.88 t/year; the removal of phosphorus was 0.38 t/year; the removal of potassium was 37.04 t/year. To reduce the inflow of surface diffuse runoff and to purify collector and drainage waters from nitrogen and phosphorus compounds, including the creation of bioplateaus and biosorption structures, it was proposed to use a set of protective measures, which will significantly reduce the biogenic load on the river flow.

**Keywords:** river basin, surface runoff, drainage waters, biogenic pollution, nitrogen, phosphorus, potassium

## 1. Introduction

Currently, the quality of water in the largest rivers in the European part of Russia continues to deteriorate. For instance, the rivers of the Volga basin are in an unsatisfactory condition [1–3]. Over the past decade, the Volga River and its tributaries have been characterised as dirty, despite decreased wastewater discharge and increased efficiency of purification of controlled discharge from point sources due to the construction of modern treatment facilities. For 1990–2012, the discharge of polluted waters decreased almost 3.5 times; the content of oil products decreased six times; the content of sulphates decreased 19.3 times; the content of chlorides decreased 3.4 times; the content of zinc decreased 15 times; the content of copper decreased 17.8 times; the content of nitrogen decreased 3.5 times; the content of phosphorus decreased 7.2 times; total biochemical oxygen consumption decreased 7.6 times [4, 5]. However, water quality in river systems and reservoirs did not improve as expected, especially in small tributaries of the Volga River. Along with the discharge of wastewater from industrial, municipal, and other enterprises, the river network is fuelled by uncontrolled diffuse runoff from the catchment area, which according to many authors, significantly worsens the quality of water in Volga [5].

The Upper Volga basin is characterised by excessive moisture content; precipitation averages 600–700 mm per year and significantly prevails over evaporation (425–475 mm). The abundance of precipitation and high snow cover lead to the formation of surface runoff, especially during the period of snow melting when solid and liquid runoffs enter the river network. This is due to erosion, leaching, and dissolution. During erosion, suspended soil particles are mainly removed with the sorption of nutrients, in particular phosphorus, on them, whereas dissolved chemicals are sorbed during dissolution and leaching. The main factors influencing the formation of surface runoff, its quality, and removal of nutrients are climate, terrain, soil surface condition, and migration capacity of nutrients. Depending on the soil type, the amount and nature of precipitation, type of plants, dose of fertilisation on one hectare of arable land, removal can be up to 80 kg of nitrate nitrogen, 3 kg of phosphorus, and 60 kg of potassium per year [6].

The abundance of precipitation requires drainage reclamation. This zone of agricultural lands is characterised by the use of horizontal drainage. As a result, drainage runoff, which is formed along with surface runoff, is directly or indirectly discharged into the river network. Through drainage waters, water bodies are filled with organic matter, residues of mineral fertilisers, and individual ions of chemical elements [7–9]. This leads to pollution of river waters and eutrophication of water bodies. Numerous studies have established that the removal of salts from drained mineral soils depends on many natural and economic indicators such as soil type, its granulometric composition, saturation with bases, the use of mineral and organic fertilisers, agricultural practices, the composition of crops, and so on. Thus, in loamy soils, the concentration of nitrogen in drainage waters varies from 5 mg/dm$^3$ to 91 mg/dm$^3$; the concentration of phosphorus varies from 0.4 mg/dm$^3$ to 0.5 mg/dm$^3$; the concentration of potassium varies from 2 mg/dm$^3$ to 10 mg/dm$^3$; the concentration of calcium varies from 61 mg/dm$^3$ to 107 mg/dm$^3$; and the concentration of magnesium varies from 21 mg/dm$^3$ to 28 mg/dm$^3$. This corresponds to the nitrogen removal of 1.4–4.1 kg/ha, phosphorus removal of up to 1 kg/ha, potassium removal of 3–12 kg/ha, calcium removal of 20–147 kg/ha, magnesium removal of 10–76 kg/ha. The concentrations of biogenic substances such as nitrogen (2.0–121.0 mg/dm$^3$), phosphorus (0.2–0.3 mg/dm$^3$), potassium (0.2–14.0 mg/dm$^3$), calcium (53–74 mg/dm$^3$), and magnesium (13–58 mg/dm$^3$) were defined in soils of lighter granulometric composition [10–14].

Drainage of the floodplain lands leads to an increased removal of nutrients directly into the river due to surface runoff and discharge of drainage waters. Thus, the mean annual nitrogen removal from agricultural lands of the floodplain lands of the Ryazan region in the Oka river, a tributary of the Volga, was 23.9 kg/ha. The concentrations of nitrates, nitrites, and ammonium were 13.7 kg/ha, 1.6 kg/ha, and 8.6 kg/ha respectively [10]. Discharge of drainage water from drainage systems causes a surge in the concentrations of nutrients and minerals in river water. From the drained floodplain of the Yakhroma river (a tributary of the Volga of the third order), the discharge of drainage water caused an increase in the concentration of ammonium ions in the river water, exceeding the $MPC_{fish}$ on average more than 2–11.5 times, so that the permissible values for the mesotrophic level increased from 4.7 to 77 times. In the summer period, an excess of the standard values of ammonia by 1.4–4 times downstream from the discharge of drainage waters was registered [15].

Thus, small rivers of the Upper Volga basin are recipients of diffuse runoff from catchments and transport impurities directly to the Volga. It seems relevant to assess the role of diffuse runoff in the general pollution of river water. The purpose of the work was to assess the diffuse runoff from the drainage basin of a small river with agricultural land in the drainage basin and to substantiate measures to protect river waters from pollution.

## 2. Selection of the study object and characteristics of its natural and climatic conditions

The study object was the drainage basin of the Yakhroma river located in the Moscow region with a catchment area of 1437 km$^2$ (**Figure 1**). The choice of the object was preconditioned by its location, a significant area of agricultural land and

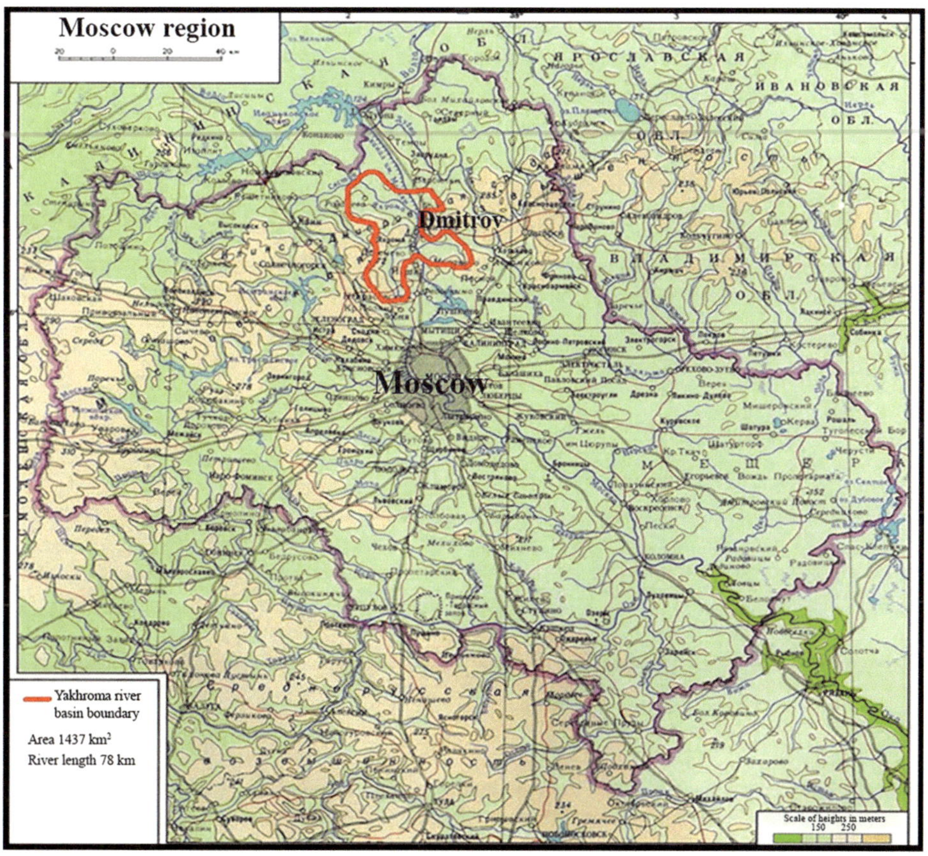


**Figure 1.**
*Physical and geographical map of the Moscow region with the allocation of the basin of the Yakhroma river.*

a large reclamation facility with an area of more than 9 th ha, on which there are 487 land use facilities and 61 drainage facilities with a total area of 26,654 ha. The studied basin comprises 249 settlements, the largest of which is the city of Dmitrov with a population of more than 68 thousand people.

Yakhroma is a third-order tributary of the Volga that flows into the Sestra river. The length of the river is 78 km [16, 17]. In the upper and partly in the middle reaches, right down to the Yakhroma reservoir, the regime of the river is natural. In the middle reaches, downstream from Dmitrov, there is a vast floodplain area, which is intensively used for agricultural production. It represents an irrigation and drainage system for bilateral regulation of the water regime of soils. Part of the area operates in polder mode. The Levyy Nagornyy canal with a system of reclamation canals built along the root bank is watered by tributaries and runoffs from settlements, which flow down to Yakhroma through a hydraulic network. Within the reclaimed massif, the Yakhroma river is canalised with its banks reinforced with dams. In the lower reach, the river flows in a low-lying area; its bed is characterised by great tortuosity (**Figure 2**).

The drainage network on the reclaimed massif is made in the form of a closed horizontal drainage 0.8–1.2 m deep with distances between drains from 12 m to 40 m. The drains flow into closed collectors, the runoff from which is discharged directly into the river through the open network. Water is taken in from Yakhroma and supplied to the irrigation network by means of mobile and stationary pumping stations. Irrigation technology is represented mainly by hose-reel sprinklers. Vegetable crops are grown on the floodplain using intensive technologies with the introduction of high doses of fertilisers and the use of various agrochemicals to combat pests and weeds. This increased load led to pollution of soil and river water with biogenic substances, mineral salts, heavy metals, and pesticides, which enter the

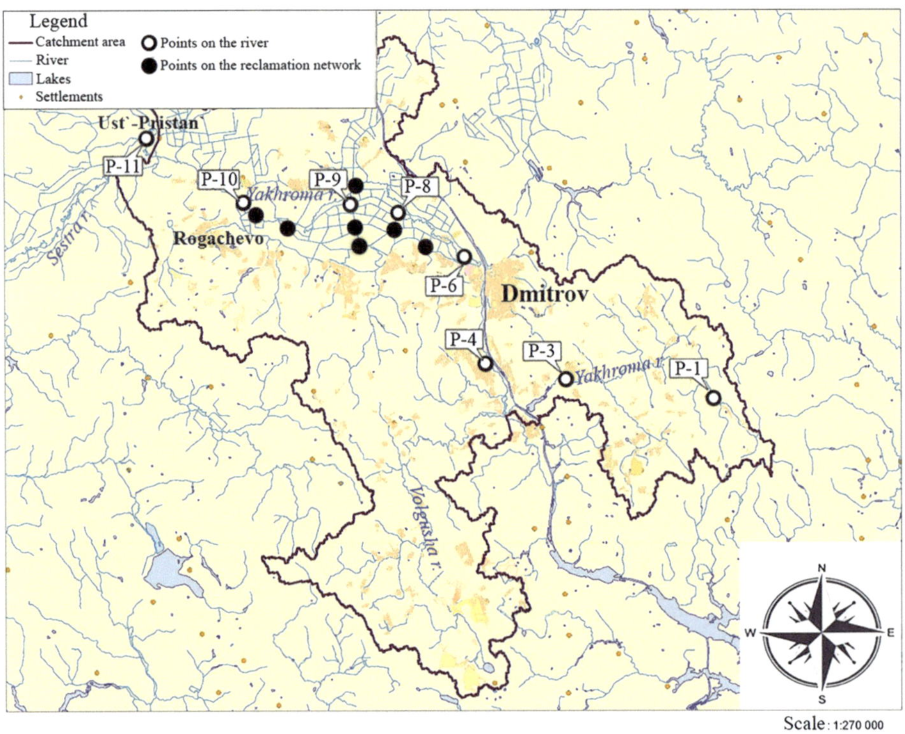


**Figure 2.**
*Map of the drainage area of the Yachroma river basin and location of observation points.*

Yakhroma river. The river serves as a drainage of flood runoff and waters from reclaimed lands of the floodplain massif.

The hydrological regime of Yakhroma is of the Eastern European type, which is characterised by high floods, low water level in the summer and winter drought periods, and increased runoff in autumn. The maximum flood discharge is, on average 10–20 times higher than the mean annual discharge [16, 17]. The share of snow supply is over 60%. Groundwater plays a significant role in nutrition.

The climate is temperate continental with frosty snowy winters and relatively humid and warm summers. According to the Dmitrov weather station, for the period 2005–2019, the mean annual air temperature was 4.9°C. The warm period with positive mean daily temperatures lasts an average of 210 days a year. The duration of the winter period with stable snow cover is more than 140 days a year. It lasts from November to mid-April. Snow height averages 55 cm. The mean annual precipitation over the past 16 years is 676 mm. Up to 70% of the annual amount of precipitation falls during the warm period from April to October. The autumn of 2019 was abnormally humid, whereas the winter of 2019–2020 was abnormally warm. Long-term dynamics of climatic conditions indicators is shown in **Figures 3** and **4**.

The soils are sod-medium podzolic, varying from gleyic to gley, light loamy sandy. The soils are depleted in organic matter (1.8–3.0%), poorly provided with mobile forms of phosphorus and potassium. Chemical elements migrate in an acidic

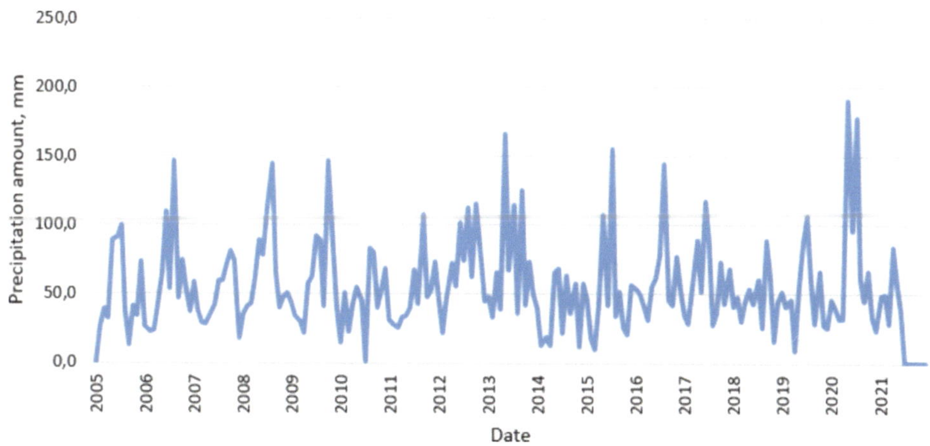


**Figure 3.**
*Chronological graph of monthly precipitation norms according to the Dmitrov meteorological station.*

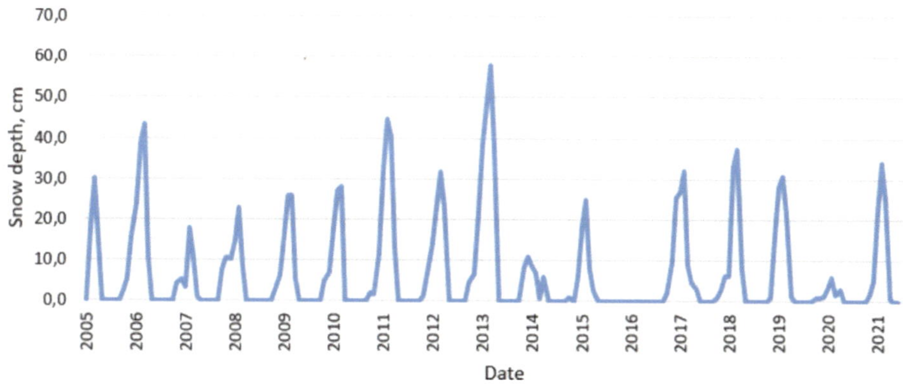


**Figure 4.**
*Chronological graph of the snow cover height according to the Dmitrov meteorological station.*

environment that leads to leaching and removal of many trace elements outside the soil profile. The valleys of the Yakhroma river are dominated by floodplain soils that form during floods, when dark, humus-rich, shallow sediments are deposited on the banks along with alluvium. The most widespread soils in the Yakhroma lowland are silt-peat, humus-peat, and humus-peaty ferruginous-carbonate gleyed. Long-term cultivation of reclaimed soils using intensive technologies with high doses of mineral fertilisers led to the secondary pollution of organogenic soil profiles with heavy metals [18].

## 3. Study methodology

The study methodology included reconnaissance along the entire length of the river with observation of the watercourse state. Observation and water sampling points were defined by various pollution sources in the catchment (**Figure 2**). Along the longitudinal profile of the river were selected eight sampling points: P-1 and P-3 points above the Yakhroma reservoir, where the anthropogenic load is not intense – the area comprises rural settlements, horticultural associations, and farmland. Point P-4 is located below the reservoir, in the zone of influence of the highway and cities; point P-6 is located at the beginning of the reclaimed massif of the Yakhroma floodplain, below the sewage canal of the city of Dmitrov; points P-8 and P-9 are located in the central part of the massif; P-10 is located at the exit from the massif; P-11 is located in the mouth part, near the Ust'-Pristan' settlement. Drainage runoff was studied in an open reclamation network and along the runoff from closed collectors; studies were also conducted on the flow of water into the reclamation array along tributaries and flood waters. The electrical conductivity, water temperature, and pH value were determined directly in situ using portable devices WTW's Cond 340i/SET and pH 330i/SET and HANNA instruments' conductometer HI 8733. In the samples taken for analysis in the laboratory, the content of potassium was determined by potentiometric methods; the content of nitrites, ammonium, and phosphates was determined by calorimetric methods. The content of metals and individual chemical elements in the Yakhroma river water and drainage canals was analysed by the spectrometric method of atomic emission with inductively coupled plasma (ICP) at the Engler-Bunt-Institute of the University of Karlsruhe in Germany (DVGW-Forshungsstelle am Engler-Bunte-Institut der Universheität (TH). The assessment of the quality of drainage runoff and its impact on the waters of the Yakhroma river involved the use of detailed studies conducted on the Yakhroma floodplain in different years by the professors of the All-Russia Research Institute of Hydraulic Engineering and Land Reclamation of A. N. Kostyakov such as Trifonov [19], Strelbitskaya [15], and Yashin [18, 20].

The removal of nutrients with surface and drainage runoff into the Yakhroma river was calculated taking into account cadastral and slope maps compiled using GIS technology for 1309 agricultural plots, 96 of which are classified into a separate group as reclaimed. For this, the Yakhroma river basin along the slopes of the terrain was zoned with identification of four large zones: zone 1 – weak flush (slope $< 0.01$), zone 2 – moderate flush (slope from 0.01 to 0.05), zone 3 – strong flush (slope $> 0.05$), zone 4 – reclaimed territory (floodplain slopeless massif) (**Figure 5**).

The zone of weak erosion, with a slope of less than 0.01, included 427 agricultural plots with a total area of 10,333.55 ha; five sections located on an area of 66.46 ha with a slope of more than 0.005 were brought together into the strong flush zone. When overlaying the zonal slope map on the soil and cadastral maps, the prevailing soil types were adopted for each agricultural plot and the total soil areas were determined by zones.

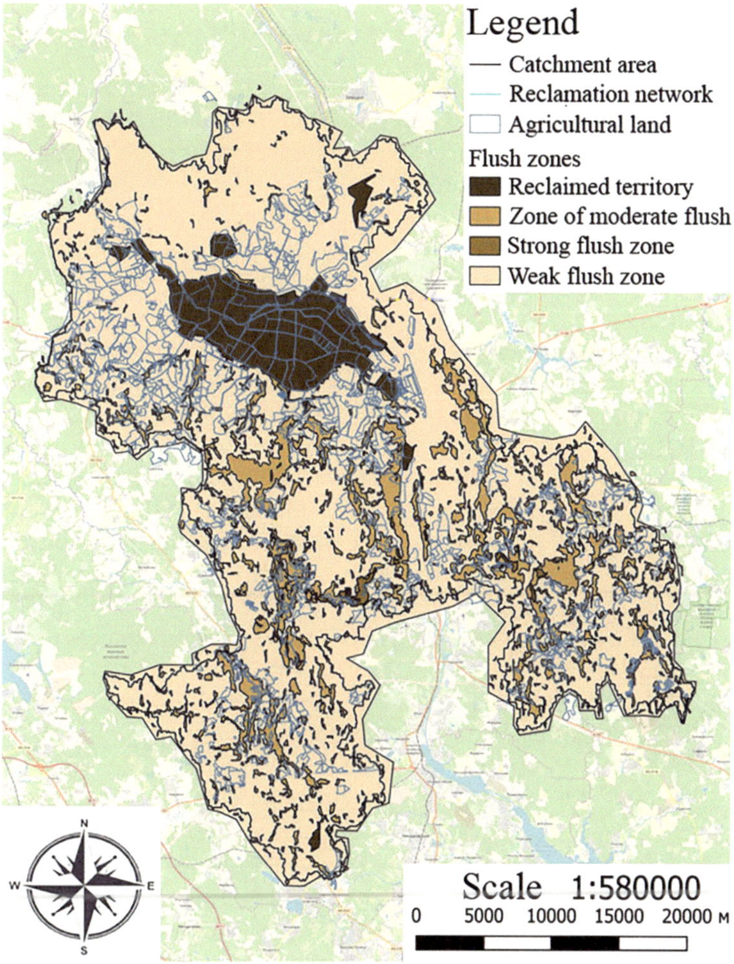


**Figure 5.**
*Map of the distribution of agricultural plots by flush zones.*

When calculating the diffuse runoff, it was assumed that potatoes are mainly grown in the weak flush zone; the moderate flush zone with sod-podzolic soils are dominated by cereals; the zone with peat soils is dominated by potatoes; in the strong flush zone, grain crops are grown on slope lands, whereas vegetables and partly potatoes are grown on reclaimed lands [21].

## 4. Calculation of the amount of surface and drainage water and the removal of nutrients into the river network

The amount of the surface runoff from the drained territory was estimated by the runoff coefficient *(σ)* depending on the amount of precipitation for a particular subzone of reclamation [22] using the formula:

$$\sigma = W_{surf.runoff} / W_{precip} \tag{1}$$

where $W_{surf.runoff}$ stands for the amount of surface runoff, $m^3$, $W_{precip}$ is the mean annual precipitation, $m^3$, determined by the formula:

$$W_{precip} = 10 \cdot H^{precip}_{av.annual} \cdot F, m^3 \quad (2)$$

where $H_{av.\ annual}{}^{precip}$ is mean annual precipitation for spring and autumn, mm; F is the water collection area, ha. It was taken into account that the surface runoff from adjacent territories enters the water intake. A river or pond having a discharge into the river was considered a water intake.

The amount of runoff was determined by the formula:

$$W_{surf.runoff} = K_{runoff} \cdot \sigma \cdot 10 \cdot H^{precip}_{av.annual} \cdot F, m^3 \quad (3)$$

where $K_{runoff}$ is the correction factor for the runoff of artificially drained territory. Other symbols are given above. The calculation results were compared with the data of field studies for the zone under consideration and, if necessary, were corrected [19, 20].

The flush of pollutants by surface runoff, including solid runoff, was determined by the dependencies outlined in the regulatory document [12]. The estimated dependencies take into account almost all sources of biogenic inputs into the soil, including the specific composition of mineral and organic fertilisers applied, the content of dissolved and absorbed biogenic substances in the soil by using appropriate correction factors. The residual amount of biogenic substances in the soil after being consumed from the soil by the crop was also taken into account.

The annual flush of absorbed and dissolved nitrogen by surface runoff ($Flush_N{}^{solid\ runoff}$) was calculated by the formula [12]:

$$\begin{aligned} Washout_N^{solid\ runoff} &= \omega \cdot \left(K_2N_y + 0.002N_O + 0.66N_n + N_{total}\right) \\ &+ \gamma\left(K_1N_y + 0.002N_O + 0,07N_n\right), kg/ha \end{aligned} \quad (4)$$

where $K_1$ is the coefficient determining the residual amount of mobile nitrogen forms of mineral fertilisers after consumption by agricultural plants (for ammonium nitrate – 0.02, ammonium sulphate – 0.03, ammonium chloride – 0.06); $K_2$ is the coefficient determining the amount of nitrogen fixed in soil and absorbed by soil microorganisms from fertilisers (for ammonium nitrate – 0.65, for ammonium sulphate – 0.35, for sodium nitrate – 0.18, for lime ammonium nitrate – 0.065); $N_y$ and $N_o$ stand for the rate of application of mineral (y) and organic (o) fertilisers, respectively, kg/ha; $N_n$ and $N_{total}$ stand for the content of mineral (n) and total nitrogen in the arable layer of soil (taken according to the survey, in this case for sod-podzolic gley soils $N_n$ = 4000 kg/ha, $N_{total}$ = 66 kg/ha, for sod-podzolic loamy soils $N_n$ = 5800 kg/ha, $N_{total}$ = 128 kg/ha, for grey forest soil $N_n$ = 5400 kg/ha, $N_{total}$ = 81 kg/ha, for leached chernozem $N_n$ = 13 kg/ha, $N_{total}$ = 195 kg/ha); $\omega$ and $\gamma$ are coefficients characterising the flush of absorbed nitrogen by solid runoff and of dissolved nitrogen from the soil surface (for peat soils $\omega$ = 3.1x10$^{-5}$, $\gamma$ = 4.3x10$^{-3}$; for sod-podzolic loamy soil $\omega$ = 7.2x10$^{-5}$, $\gamma$ = 4.8x10$^{-3}$; for grey forest soil $\omega$ = 1.8x 10$^{-4}$, $\gamma$ = 1.4x10$^{-2}$; for leached podzolised chernozem $\omega$ = 4x10$^{-5}$, $\gamma$ = 2.4x10$^{-2}$).

The flush of absorbed phosphorus with solid runoff ($B_P{}^{solid\ runoff}$) over the year was determined as follows:

$$Washout_P^{solid\ runoff} = \omega \cdot \left(n_2P_y + n_3P_o + n_4P_n + P_{total}\right), kg/ha \quad (5)$$

where $P_y$ and $P_o$ stand for the rate of application of mineral and organic fertilisers, respectively, kg/ha; $P_n$ and $P_{total}$ stand for the content of mineral and total phosphorus in the arable layer of soil (for sod-podzolic gley soils $P_n$ = 300 kg/ha, $P_{total}$ = 1820 kg/ha; for sod-podzolic loamy soils $P_n$ = 210 kg/ha,

$P_{total}$ = 3380 kg/ha; for grey forest soil $P_n$ = 420 kg/ha, $P_{total}$ = 3600 kg/ha; for leached chernozem $P_n$ = 260 kg/ha, $P_{total}$ = 3900 kg/ha); $n_2$, $n_3$, and $n_4$ are coefficients characterising the residual amount of phosphorus in mineral, organic fertilisers, and soil respectively ($n_2$ for light soils – 0.8; for heavy soils – 0.26; for peat soils – 0.32; $n_3$ = 0.0014, 0.0004, and 0.0005; $n_4$ = 0.85, 0.28, and 0.34).

The annual flush of absorbed and dissolved potassium by surface runoff ($Flush_K^{surf.runoff}$) was calculated by the formula:

$$\begin{aligned} Washout_K^{surf.runoff} &= \omega \cdot (0.2K_y + 0.0012K_{total} + 0.008K_{total} + K_{total}) \\ &+ \gamma[((0,2K_y + 0.0012K_{total} + 0.008K_{total} + K_{total}) \cdot 0.018)] \end{aligned} \quad (6)$$

where $K_y$ is the rate of application of the mineral fertiliser, kg/ha; $K_{total}$ is the total amount of potassium in the arable layer of soil, kg/ha (for sod-podzolic gley soils $K_{total}$ = 50,000 kg/ha, for sod-podzolic loamy soils $K_{total}$ = 58,000 kg/ha, for grey forest soil $K_{total}$ = 50,600 kg/ha, for leached chernozem $K_{total}$ = 51,250 kg/ha).

The drainage flow ($W_{dr}$) was calculated based on the known dependencies of the mean annual module drainage flow:

$$W_{dr} = \frac{q \cdot F \cdot t}{1000}, m^3 \quad (7)$$

where q is the mean annual module of drainage flow, L/s/ha; F is the area of the drained area, ha; t is the number of seconds in a year, s.

A correction factor for the flush of biogenic substances for the long-term mean annual water content was introduced to determine the flush of biogenic substances with drainage runoff.

Nitrogen annual flush by drainage runoff ($Flush_N^{dr}$) is determined by the formula:

$$Washout_N^{dr} = \frac{(K_1 \cdot N_y + 0,0002N_0 + 0,007N_n)W_{dr}}{W_{limit} + W_{dr}}, kg/ha \quad (8)$$

where $W_{limit}$ is moisture reserve in the considered soil layer to the groundwater depth or to the depth of drainage at maximum moisture capacity of soil, $m^3$/ha (for peat soils $W_{limit}$ = 4500 $m^3$/ha, for sod–podzolic loamy soils $W_{limit}$ = 2682 $m^3$/ha, for grey forest soil $W_{limit}$ = 2138 $m^3$/ha, for leached chernozem $W_{limit}$ = 2765 $m^3$/ha);

See other symbols above.

The annual flush of dissolved phosphorus by drainage runoff ($Flush_P^{dr}$) is determined by the formula:

$$Washout_P^{dr} = \frac{n_1 W_{arable}^{limit} \cdot W_{dr}}{W_{limit} + W_{dr}}, kg/ha \quad (9)$$

where $n_1$– the value characterising the content of dissolved phosphorus in the soil (for light soils it is 0.002; for heavy soils it is 0.00017; for peat soils it is 0.0015); $W_{arable}^{limit}$ is moisture content in the topsoil, $m^3$/ha (for peat soils $W_{limit}$ = 1350 $m^3$/ha, for sod–podzolic loamy soils $W_{limit}$ = 537 $m^3$/ha, for grey forest soil $W_{limit}$ = 428 $m^3$/ha, for leached chernozem $W_{limit}$ = 553 $m^3$/ha).

The annual flush of dissolved potassium by drainage runoff ($Flush_K^{dr}$) is determined by the formula:

$$Washout_K^{dr} = \frac{[(0,2K_y + 0,0012K_0 + 0,008K_{total})0,018]W_{dr}}{W_{limit} + W_{dr}}, kg/ha \quad (10)$$

where $K_o$ is the rate of application of organic fertiliser, kg/ha,

The total volume of biogenic compounds flush was determined by the following formula:

$$B^{total} = \sum B_i^{dr} + \sum B_i^{surf} = B_N^{dr} + B_P^{dr} + B_K^{dr} + B_N^{surf} + B_P^{surf} + B_K^{surf}, \text{kg/ha} \quad (11)$$

where $B_i^{dr}$, $B_i^{surf}$ stand for the flush of i biogenic element (nitrogen, phosphorus, potassium) by the drainage and surface runoff respectively.

## 5. Hydrochemical studies of water quality in the Yakhroma river basin

The purpose of hydrochemical studies conducted since 2001was to establish and localise sources of water pollution along the longitudinal profile of the river. They showed that over the entire observation period, the salinity of water in the river fluctuated within 80–500 mg/dm$^3$ and increased from source to mouth due to inflow of wastewater from the cities of Yakhroma and Dmitrov and surface runoff of tributaries and diffuse runoff from the catchment area. The data on the content of biogenic pollutants is taken from observations conducted in 2019–2020. In the spring period (samples were taken on April 12, 2019), the chemical composition of river water was determined mainly by the quality of melt water from the slopes and along the tributaries and was characterised by relatively low values of the content of dissolved salts. It was found that along the longitudinal profile of the river, the electrical conductivity naturally increased due to diffuse runoff and point sources from 56 μS/cm in the upper reaches to 229 μS/cm in the lower reaches (**Figure 6**). The highest value is confined to the site where the river receives wastewater from the city of Dmitrov and drainage water from the reclaimed massif, which are characterised by increased electrical conductivity (P-6–P-10). The electrical conductivity of the tributaries entering the Levyy Nagornyy canal varies from 285 μS/cm to 591 μS/cm; in the open drainage network it varies from 352 μS/cm to 1170 μS/cm. At the same time, increased values are typical for drainage canals with minimal flow rates.

The distribution of electrical conductivity values along the longitudinal profile of the river in the summer (July 15, 2020) and autumn (September 30, 2020)

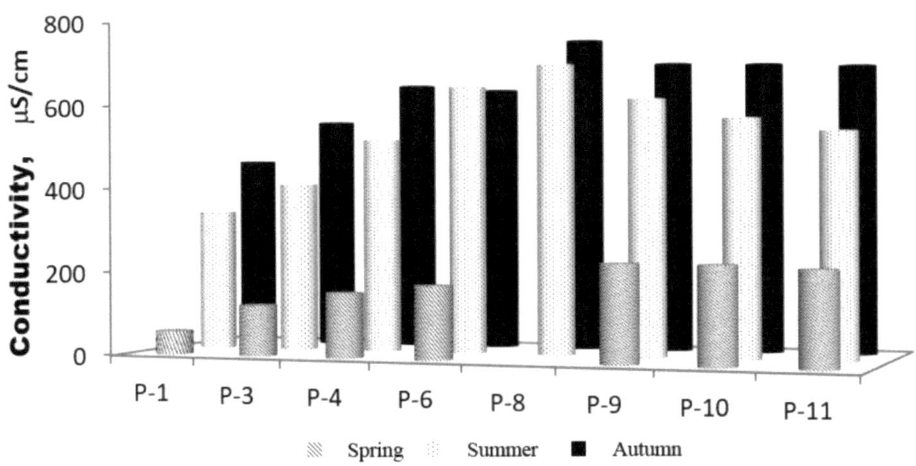


**Figure 6.**
*Change in electrical conductivity (μS/cm) of water along the longitudinal profile of the Yakhroma river by season.*

periods is similar. These seasons are characterised by higher values reaching 695–766 μS/cm. The maximum values are observed within the reclaimed massif, which is associated with drainage waters flowing into the river. The ratios of the electrical conductivity values of the water of the Yakhroma river by seasons in the sections "entry" to the reclaimed massif and "exit" from it showed that due to drainage runoff and unloading of groundwater with an increased concentration of salts, the salinity of the river runoff increases in winter–spring periods – mineral salts are withdrawn into the river by runoff from agricultural lands. In summer, due to irrigation and decreased salinity of the drainage runoff, electrical conductivity at the outlet from the massif decreases. In the autumn period, due to precipitation, these processes slow down so that the values of electrical conductivity become almost similar. The values of the electrical conductivity of almost all hydrospheric components within the reclaimed massif, tributaries from the sides of the valley (520–870 μS/cm) and pressure waters (640–650 μS/cm) show that the maximum values of electrical conductivity reaching 1100–1300 μS/cm are characteristic of the drainage runoff in autumn and winter. This is probably due to partial flush of mineral fertilisers by the surface runoff and flow of residual amounts into deep soil horizons and groundwater.

**Figure** 7 shows the distribution of nutrient concentrations in river water along the longitudinal profile from the source to the mouth of the Yakhroma river. The nutrient content in river water gradually increases from the source to the beginning of the reclaimed Yakhroma floodplain massif (point P-6) with some fluctuation within the land reclamation massif and a decrease towards the river mouth. A sharp increase is observed at point P-6. Increased concentrations of phosphates and ammonium are characteristic of the summer and autumn periods and are confined to the reclaimed massif of the Yakhroma floodplain (P-8–P-10). The surge in the concentrations of phosphates and ammonium nitrogen is confined to the beginning of the floodplain massif (R-6) and is caused by the influence of the discharge of insufficiently treated urban wastewater. On reclaimed lands, the content of phosphates in the reclamation network varies, as a rule, in the range of 0.06–0.54 mg/dm$^3$, ammonium nitrogen content varies between 1.3 and 4.5 mg/dm$^3$ and 11.03 mg/dm$^3$, which significantly exceeds the fishery standards. The potassium content varies over a wide range – from 1.7 to 33.7 mg/dm$^3$, while the most frequent values fall into the range 2.0–10.8 mg/dm$^3$. This is confirmed by detailed studies conducted earlier by A.V. Trifonov [19], who showed that the annual drainage removal of potassium oxide is 8 kg/ha. Drainage waters also contain ions of calcium, magnesium, iron, nitrogen, sulphur, chloride, potassium, phosphorus, and silicon. According to E.B. Strelbitskaya's studies [15], the inflow of drainage water from the reclaimed massif increased the concentration of ammonium ions in the river water in the area below the discharges from the drainage system, exceeding the standards for fishery reservoirs on average by more than 2–11.5 times.

The results received in 2019–2020 are confirmed by the study conducted by V.M. Yashin in 2001–2005 on the reclaimed massif of the Yakhroma floodplain [18]. The pH varies from 6.0 to 8.1 with the most frequent values in the range from 7.0 to 7.7. There are no definite patterns in pH changes for various water bodies. Artesian (0.4–0.8 mg/dm$^3$) and drainage waters at the mouths of closed drains and collectors (1–2 mg/dm$^3$) are characterised by the lowest values of dissolved oxygen. In open reclamation canals, the content of dissolved oxygen does not reach the standard (6.0 mg/dm$^3$) level. The maximum concentrations of biogenic pollutants are typical for the Levyy Nagornyy canal, which receives water from tributaries from the left side of the valley, including groundwater and partially wastewater from rural settlements.

The pollution of drainage and river waters with heavy metals is characterised by the data in **Table 1**. The water contains a wide range of dissolved metals. On the

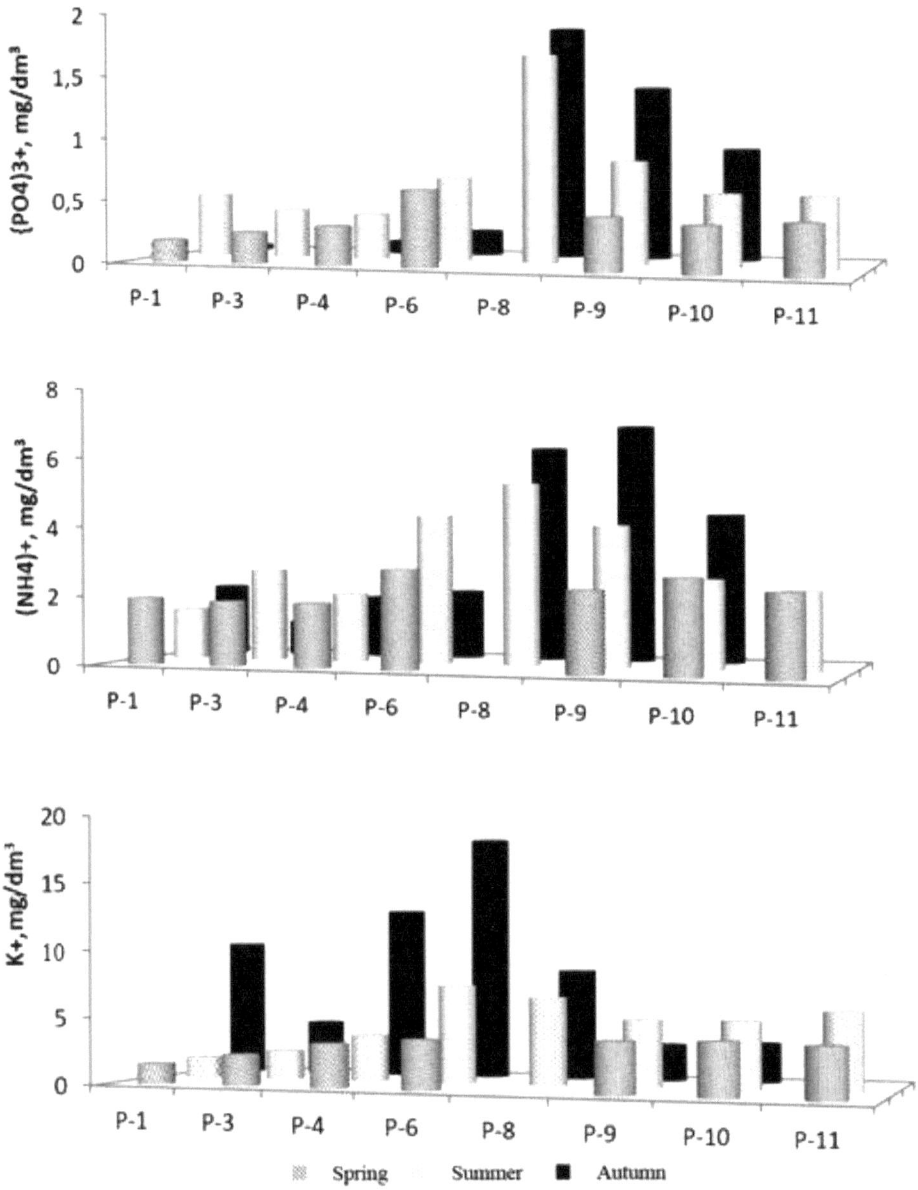


**Figure 7.**
*Distribution of the content of nutrients along the length of the longitudinal profile of the Achromat river by seasons.*

basis of the concentrations of metals drainage, river waters are divided into three groups: more than 1000 μg/dm$^3$ (calcium, magnesium, sodium, potassium and silicon for river waters and iron and strontium for drainage waters), 100–999 μg/dm$^3$ (aluminium, barium, phosphorus, manganese, and iron for drainage waters and strontium for river waters), and up to 99 μg/dm$^3$ (heavy metals such as cobalt, copper, nickel, lead, zinc). The concentrations of metals, except iron and strontium, in the drainage runoff exceed their content in river water, which indicates potential danger of river water pollution.

The concentrations of organochlorine (according to AEX) compounds in drainage and river waters were 23 μg/dm$^3$ and 8.2 μg/dm$^3$ respectively.

| Metal | Open drainage | Closed drainage | | Yakhroma river, output from massive |
|---|---|---|---|---|
| | | KYa-26-2 | KYa 26-2-12 | |
| Ca | 111,700 | 150,820 | 162,080 | 92,830 |
| Mg | 24,835 | 30,030 | 50,455 | 24,665 |
| Na | 13,380 | 8964 | 8435 | 18,890 |
| K | 2404 | 2025 | 2950 | 5170 |
| Fe | **254** | **3230** | 98 | **254** |
| Si | 983 | 9000 | 12,360 | 8260 |
| Sr | **534** | **614** | **3530** | **573** |
| Al | 13 | **45** | 15 | 26 |
| Ba | **87** | **125** | **115** | 54 |
| Mn | **200** | **500** | **580** | **48** |
| Co | 1 | 2 | 1 | <1 |
| Ni | <2 | **12** | **50** | 6 |
| P | | **42** | **199** | **271** |
| Pb | <5 | <5 | **99** | <5 |
| Cu | 2 | 5 | **3** | <2 |
| Zn | 3 | 5 | **100** | **13** |
| V | 2 | **4** | **50** | **4** |

*Note: metal concentrations exceeding the standards are highlighted in bold.*

**Table 1.**
*Content of metals in drainage runoff and river water, (μg/dm³).*

Thus, the study revealed increased concentrations of pollutants in the river water within the reclaimed massif of the Yakhroma floodplain. This confirms the influx of uncontrolled diffuse runoff into the river network from the catchment area.

## 6. Assessment of diffuse runoff from the catchment area of the Yakhroma river

To assess the diffuse runoff, the volume of surface water was calculated for the identified zones of flush and drainage water from the reclaimed massif (**Figure 5**). The average long-term module of surface runoff depends on the slope angle. In the zones of weak, moderate, and strong flushes, it equals to 0.013 $dm^3/s/ha$, 0.017 $dm^3/s/ha$, and 0.029 $dm^3/s/ha$ respectively. On the reclaimed massif, where drainage is used, the module of surface runoff is less than 0.005 $dm^3/s/ha$; the calculated mean vegetation module of drainage runoff is 0.052 $dm^3/s/ha$; the maximum module during the spring snowmelt period reaches 1 $dm^3/s/ha$. The mean annual amount of surface and drainage runoff was calculated according to formulas (3) and (7) for all considered zones (**Table 2**).

As can be seen from the table, surface runoff is 12.4 mln $m^3$/year, while drainage runoff is 15.9 mln $m^3$/year. The first is formed on an area of 22.6 th ha, while the second is concentrated on a floodplain with an area of 9.7 th ha. Removal of nutrients (NPK), surface, and drainage runoff are calculated according to dependencies (4–6) and (8–10). Calculations showed that the annual surface removal of

| Zone name | Zone 1, weak flush, slope <0.01 | Zone 2, moderate flush, slope from 0.01 to 0.05 | Zone 3, strong flush, slope > 0.05 | Zone IV, floodplain massif |
|---|---|---|---|---|
| Area, ha | 10333.55 | 12,233.99 | 66.46 | 9730.80 |
| Drainage | | | | 15,957.26 |
| Surface runoff | 4173.72 | 6551.32 | 61.34 | 1572.11 |
| Total flow, th $m^3$/year | 28,315.75 | | | |

**Table 2.**
*The results of calculating the amount of surface and drainage runoff (thousand $m^3$/year) from the drainage basin of the Yakhroma river.*

nitrogen from agricultural land varies from 0.51 kg/ha to 1.09 kg/ha, while drainage removal is 9.96 kg/ha (**Table 3**).

The largest nitrogen removal is noted with drainage waters; its average concentration in water reaches 6 mg/$dm^3$; the predominant form of nitrogen in drainage water is nitrate (97% of the removed total mineral nitrogen), the content of which in drainage water according to data [23] can vary within the range of 0.42–9.66 mg/$dm^3$, whereas the removal reaches 5.5–8.0 kg/ha.

The removal of phosphorus from the soil by surface and drainage waters is insignificant (does not exceed 0.1% of the phosphorus content in the soil), which is explained by its low mobility (**Table 4**).

Phosphorus practically does not enter the drainage waters; its concentration does not exceed 0.02 mg/$dm^3$ even with the introduction of phosphate fertilisers, since orthophosphoric acid is associated mainly with trivalent metals, and especially with aluminium. In the classical experiments of the Rothamsted Experimental Station on the use of phosphorus fertilisers conducted for 130 years, the

| Zone name | Zone 1 | Zone 2 | Zone 3 | Zone 4 |
|---|---|---|---|---|
| Nitrogen removal by surface runoff, t/year | 9.59 | 6.42 | 0.03 | 11.32 |
| Weighted average nitrogen removal by surface runoff, kg/ha | 0.92 | 0.51 | 0.53 | 1.09 |
| Nitrogen removal by drainage runoff, t/year | | | | 96.88 |
| Weighted average nitrogen removal by drainage runoff, kg/ha | | | | 9.96 |
| Nitrogen concentration in surface runoff | 2.30 | 0.98 | 0.49 | 7.20 |

**Table 3.**
*Results of calculating nitrogen removal from the drainage basin of the Yakhroma river by flush zones.*

| Zone name | Zone 1 | Zone 2 | Zone 3 | Zone 4 |
|---|---|---|---|---|
| Phosphorus removal by surface runoff, t/year | 2.02 | 2.41 | 0.01 | 1.62 |
| Weighted average phosphorus removal by surface runoff, kg/ha | 0.16 | 0.14 | 0.20 | 0.14 |
| Phosphorus removal by drainage runoff, t/year | | | | 0.38 |
| Weighted average removal of phosphorus by drainage runoff, kg/ha | | | | 0.04 |
| Phosphorus concentration in surface runoff, mg/$dm^3$ | 0.48 | 0.37 | 0.14 | 1.03 |
| Concentration of phosphorus in the drainage runoff, mg/$dm^3$ | | | | 0.02 |

**Table 4.**
*Results of calculating phosphorus removal by flush zones from the drainage basin of the Yakhroma river.*

| Zone name | Zone 1 | Zone 2 | Zone 3 | Zone 4 |
|---|---|---|---|---|
| Potassium removal by surface runoff, t/year | 79.67 | 100.94 | 0.52 | 61.15 |
| Weighted average potassium removal by surface runoff, kg/ha | 7.89 | 5.62 | 7.89 | 4.87 |
| Removal of potassium by drainage runoff, t/year | | | | 37.04 |
| Weighted average potassium removal by drainage runoff, kg/ha | | | | 3.81 |
| Potassium concentration in surface runoff, mg/dm$^3$ | 19.09 | 15.41 | 8.53 | 38.89 |
| Potassium concentration in drainage runoff, mg/dm$^3$ | | | | 2.32 |

**Table 5.**
*Results of calculating potassium removal from the drainage basin of the Yakhroma river by flush zones.*

concentration of phosphorus in drainage waters was not more than 0.05 mg/dm$^3$ with a phosphorus content in the soil solution of up to 0.2 mg/dm$^3$ [24].

The removal of potassium with surface and drainage waters significantly depends on its content in the soil and the application rates of highly soluble potassium fertilisers and ranges from 4.87 kg/ha to 7.89 kg/ha (**Table 5**).

In the drainage waters of soddy-gley soils, the concentration of potassium reaches 2.32 mg/dm$^3$. This is due to the introduction of a large amount of potash fertilisers on the reclaimed massif, which is confirmed by the earlier studies by M.A. Borovitskaya [25]: when potassium was applied at a dose of 170 kg/ha, the removal of potassium on soddy-podzolic loamy soil increased from 3.3 kg/ha to 9.6 kg/ha, and on sandy loam it increased from 2.2 kg/ha to 10.5 kg/ha.

The total removal of nutrients, according to calculations, amounted to almost 410.0 t/year, including removal of nitrogen compounds of 124.2 t/year, removal of phosphorus of 6.4 t/year, and removal of potassium of 279.4 t/year [26]. Drainage runoff removes 96.9 t/year of nitrogen compounds, whereas surface waters remove only 27.4 t/year, which is 3.5 times less, while surface removal of phosphorus compounds is on the contrary 6.1 t/year and drainage removal is 0.4 t/year, or 15 times less. Removal of nutrients with drainage runoff on reclaimed lands is almost two times higher than with surface runoff and amounts to 134 kg/ha/year. Dissolved nitrogen compounds predominate in the drainage runoff. Their removal varies from 7 kg/ha/year to 12 kg/ha/year; drainage removal of phosphorus compounds varies from 0.02 kg/ha/year to 0.06 kg/ha/year; potassium removal varies from 0.8 kg/ha/year to 5.7 kg/ha/year. This explains the surge in the pollution of the Yakhroma river waters within the reclaimed massif.

## 7. Development of measures to reduce diffuse pollution of the Yakhroma river

The pollution of river waters can be reduced by decreasing surface runoff and improving its quality. Low-cost organisational and economic measures reduce diffuse runoff by 20%, agrotechnical methods reduce it by 25–50%, and agro-reclamation methods reduce it by 50–75%. Organisational, economic, and agrotechnical measures are recommended on all plots of arable land of the catchment of the Yakhroma river on an area of 32,365.5 ha.

On arable land with a slope of up to 0.01 (weak flush zone) on an area of 10333.6 ha, it is recommended to conduct accelerated ploughing to transfer part of the surface runoff to soil runoff by forming shallow parallel furrows every 4–15 m, into which surface water flows from the entire enclosure. Lands with small slopes (up to 0.01) should be ploughed towards the natural surface slope, with large slopes

| Scenarios | Removal of nutrients in the current state | | | | Predicted removal of nutrients after taking the measures | | | |
|---|---|---|---|---|---|---|---|---|
| Zone name | Zone 1 | Zone 2 | Zone 3 | Zone 4 | Zone 1 | Zone 2 | Zone 3 | Zone 4 |
| Nitrogen removal, t/year | 9.59 | 6.42 | 0.03 | 108.19 | 7.67 | 3.21 | 0.00 | 26.48 |
| Phosphorus removal, t/year | 2.02 | 2.41 | 0.01 | 1.99 | 1.61 | 1.21 | 0.00 | 0.42 |
| Potassium removal, t/year | 79.67 | 100.94 | 0.52 | 98.19 | 63.73 | 50.47 | 0.00 | 21.49 |
| Total removal of nutrients, t/year | 91.27 | 109.77 | 0.56 | 208.38 | 73.02 | 54.89 | 0.00 | 48.39 |
| Total removal of nutrients, t/year | 409.984 | | | | 176.294 | | | |

**Table 6.**
*Estimated values of nutrient removal during water protection measures.*

at a certain angle to the surface slope so that the slope of the furrows does not exceed 0.01. The recommended distance between the exit furrows is 80–140 m. These techniques will reduce the surface runoff by up to 50%.

In the zone of moderate erosion on an area of 12,234 ha, it is recommended to conduct surface ridging or deep ridge ploughing. This leads to the formation of dense network of furrows, the water from which is diverted to the collectors along the drainage furrows, which reduce the diffuse runoff by 50–75%.

The strong flush zone occupy an area of 66.5 ha. To regulate surface runoff, it is recommended to create anti-erosion hydraulic structures, which are earthen embankments, that is, terraces that allow partial surface runoff without destroying the soil in cases of rainfall in excess of the calculated value.

On the reclaimed massif of the Yakhroma floodplain under study on an area of 9731 ha, where the drainage runoff is discharged directly into the river network, it is recommended to build a diversion canal with a bed bioplateau inside it, downstream from which the treated runoff is discharged into the Yakhroma river. For additional treatment of the drainage runoff, the bioplateau is planted with higher aquatic vegetation in alternating strips of 5–10 m along the width of the watercourse serving as a barrier to the incoming pollutants. The flood is passed through the retaining structure. At a depth of from 0.8 m to 1.2 m in structures, it is recommended to plant *Phragmítes austrális* and *Ceratophýllum demérsum* to absorb nutrients and *Schoenoplēctus lacūstris*, *Elodeacanadensis*, and *Ceratophýllum demérsum* to reduce the concentration of heavy metals, phenol, and pesticides. The bioplateau purifies the drainage runoff from biogenic pollutants by 55–85%.

A possible reduction in the removal of pollutants into water bodies of the Yakhroma river was also calculated when taking the above measures. Also the removal of nutrients such as nitrogen (37.36 t), phosphorus (3.24 t), and potassium (135.7 t) was determined. The total intake of nutrients (176.3 t) allows reducing the diffuse load on the water body by more than 50% (**Table 6**).

Thus, for the given catchment area, the recommended set of measures will ensure a decrease in diffuse runoff formed as a result of agricultural activities.

## 8. Conclusions

1. To assess the possible diffuse pollution of water bodies in the process of agricultural production, the catchment area of the small Yakhroma river was selected. The conducted studies made it possible to comprehensively consider

the formation of diffuse runoff from the surface of agricultural fields and drainage runoff from the reclaimed objects depending on climatic, soil, and organisational conditions and to estimate the volume of biogenic pollution entering the Yakhroma river.

2. Hydrochemical studies of river water quality along the Yakhroma river profile made it possible to identify pollution sources, the main of which are the point discharge of insufficiently treated municipal waters from Dmitrov and Yakhroma cities and the discharge of drainage and surface waters from the reclaimed massif of the Yakhroma river floodplain.

3. The methodological approach to the assessment of regional diffuse pollution proposed by the authors provides a solution to one of the priority tasks of environmental management and is of practical importance in assessing the pollution of any water body with biogenic substances. The study methodology is based on a comprehensive and objective analysis of the results of field studies in the region under consideration.

4. The revealed regularities of the formation of surface and drainage runoffs and the use of calculated dependences make it possible to comprehensively consider the formation of diffuse runoff from the surface of agricultural fields and drainage runoff from drainage systems depending on climatic, organisational, and economic conditions and to estimate the volume of biogenic pollution entering the water bodies of the Upper Volga.

5. It is shown that in the Yakhroma river basin the amount of surface runoff from agricultural fields is 12.4 mln $m^3$/year; the amount of drainage runoff is 15.9 mln $m^3$/year, while surface runoff is formed on an area of 22.6 th ha, and drainage is concentrated on a floodplain with an area of 9.7 th ha The total removal of nutrients, according to calculations, amounted to almost 410.0 t/year, including that of nitrogen compounds (124.2 t/year), phosphorus (6.4 t/year), and potassium (279.4 t/year). Removal of nutrients with drainage runoff on drained lands is almost 2 times higher than with surface runoff and amounts to 134 kg/ha/year. Dissolved nitrogen compounds predominate in the drainage runoff. Their removal varies from 7 kg/ha/year to 12 kg/ha/year; the removal of phosphorus compounds with drainage runoff varies from 0.02 kg/ha/year to 0.06 kg/ha/year; potassium removal varies from 0.8 kg/ha/year to 5.7 kg/ha/year.

6. Reducing and cleaning diffuse runoff will allow decreasing the diffuse load on a water body by more than 50%, protecting water bodies, and reducing risks to human life and health.

## Author details

Liudmila V. Kireicheva*, Valery M. Yashin, Ekaterina A. Lentyaeva
and Aleksey D. Timoshkin
All-Russian Research Institute of Hydraulic Engineering and Reclamation Named After A.N. Kostyakov, Moscow, Russia

*Address all correspondence to: kireychevalw@mail.ru

## References

[1] Dzhamalov RG, Safronova TI, Trofimchuk MM, et al. Average long-term features of the formation of the chemical composition and quality of the waters of the Volga basin. Proceedings of the All-Russian Scientific Conference with International Participation "Scientific Problems of Sanitation of Russian Rivers and Ways to Solve Them"; Nizhny Novgorod; 8–14 September 2019; Moscow: Studiya F1; 2019. 68–74

[2] Gremm TJ, Heidt A, Ftimmel FH. Quality of Russian rivers. Chemistry in Our Time. 2002;**4**:226–239

[3] Novoseltsev VN et al. Technogenic Pollution of River Ecosystems. Moscow: Nauchnyy Mir; 2002. p. 140

[4] State Report "The State and Protection of the Environment of the Russian Federation in 2016". Moscow: NIA-Priroda; 2017. p. 760

[5] Komarov IK. The Revival of the Volga as a Step Towards the Salvation of Russia. Moscow – Nizhny Novgorod: Ecology; 1996. p. 464 e

[6] Komparskas II. The Regime of Nutrient Leaching by Drainage Runoff on Drained Mineral Soils. Vilnius: Proceedings of Lithuanian Research Institute of Hydraulic Engineering and Land Reclamation; 1966. p. 324

[7] Shkinkis Ts.N. The Hydrological Effect of Drainage. Leningrad; Gidrometeoizdat: 1981. p. 311

[8] Fedotova ZD, Strautynya VP. Removal of Nutrients by the Drainage Runoff of Drained Soils. Jelgava: Proceedings of Lithuanian Research Institute of Hydraulic Engineering and Land Reclamation. 1969;9:43–47

[9] Khrisanov NI, Osipov GK. Eutrophication of Water Bodies. St. Petersburg; Gidrometeoizdat; 1993. p. 276

[10] Ecological Problems of the Upper Volga: Collective Monograph. Yaroslavl; Publishing House of YaSTU: 2001. p. 427

[11] Betson RP. and McMaster M. Non-point source mineral water quality model. Journal (Water Pollution Control Federation). 1975;**47**:10. 42

[12] Hock B. Water Quality Balance. Technical and Economic Water Management Series VMGT. Vol. 27. Budapest, Hungary: VIZDOK Press; 1970. p. 124

[13] Maidment DR. Handbook of Hydrology. New York, NY: McGrow-Hill Inc.; 1992. p. 87

[14] O'Brien WG. et al. Modeling Discharge and Conservative Water Quality in the Lower Kansas River Basin. Univ. of Kansas. Bull. 204, Part 3. 1972. p. 153

[15] Strelbitskaya EB, Kolomiytsev NV. Changes in the ecological state of small rivers under the influence of waste waters from drained agricultural landscapes and ways to improve it. Reclamation and Water Management. 2006;**5**:38–43

[16] Wagner BB. Rivers and Lakes of the Moscow Region. Moscow: Veche; 2006. p. 480

[17] Wundzettel MF. Ichthyofauna of small rivers and reservoirs of the northern Moscow region. ASTU Bulletin. Fisheries Series. 2012;**1**:7–14

[18] Yashin VM. Formation of the quality of drainage flow on the Yakhroma floodplain. Reclamation and Water Management. 2017;**6**:21–26

[19] Trifonov VA. Removal of Chemicals by Drainage Runoff from Drained Floodplain Soils and Its Regulation. PhD in technical sciences [thesis]. Moscow: All-Russia Research Institute of Hydraulic

Engineering and Land Reclamation of A. N. Kostyakov; 1989. p. 329

[20] Kireicheva LV, Yashin VM. Study of diffuse pollution of small rivers: Case study of the Yakhroma river. International Research Journal. 2020; **3–1**(93); 85–90. DOI: 10.23670/IRJ.2020.93.3.012

[21] Kireicheva LV, Lentyaeva EA, Timoshkin AD, Yashin VM. Assessment of diffuse pollution from agricultural areas in the Upper Volga basin and development of measures to reduce it using the case study of the Yakhroma river. Water Resources. 2020;**47.5**:523–535. DOI 10.31857/S0321059620050090

[22] Guidelines for Determining the Estimated Concentrations of Mineral and Organic Substances and Pesticides in Drainage and Surface Runoff from Reclaimed Lands. VTR–P–30–81. Moscow: Ministry of Land Reclamation and Water Management; 1981. p. 42

[23] Amberger A, Schweiger P. Migration of plant nutrients in the soil and their importance in environmentally conscious agriculture. Die Bondekultur. 1973

[24] Cooke GW. Review of the effects of agriculture on the chemical composition and quality of surface and underground waters. Agriculture and Water quality. Technical Bulletin 32. London; 1976

[25] Borovitskaya MA. Migration of elements in the local geochemical landscape of the Istra River, Moscow region. Soil Science. 1964;**11**:79–87

[26] Kireycheva LV, Lentyaeva EA, Timoshkin AD, Yashin VM. Assessment of the diffuse pollution from agricultural territories in the upper Volga basin and the development of measures to reduce it: Case study of the Yakhroma river. Water Resources. 2020;**47.5**:709–720. Doi: 10.1134/s0097807820050097

Chapter 3

# Implications of Land Use and Cover Changes on Upper River Rwizi Macro-Watershed Health in South Western Uganda

*Denis Nseka, Hosea Opedes, Frank Mugagga, Patience Ayesiga, Henry Semakula, Hannington Wasswa and Daniel Ologe*

## Abstract

The upper Rwizi river system in South Western Uganda has been severely degraded due to encroachment and unsustainable resource utilization. Little is, however, known about the link between the upper Rwizi macro-watershed health and land use/cover patterns from the spatiotemporal perspective. This study evaluated the relationship between spatiotemporal land use/ cover change and upper river Rwizi macro watershed health. Remotely sensed data was used to analyze the spatiotemporal land use and cover distribution for upper Rwizi macro watershed. The analysis was done using Landsat and Sentinel imagery datasets spanning 1990 to 2020 and 2016 to 2021 respectively. Field verification was undertaken to confirm the land use, cover types, and evaluate the implications of prevailing anthropogenic activities on the watershed health. The land use and cover characteristics in the upper Rwizi macro-watershed exhibits both highly spatial and temporal variations. By 1990, grassland as the dominant land use and cover type spanned 45% of the total study area followed by farmland at 30%. Forests, open water and settlements covered 12%, 10% and 3% respectively. Whereas grassland and forest cover has diminished drastically by 64% and 71% respectively, settlements and farmland have increased tremendously by 79% and 50% respectively between 1990 and 2020. The hillslope hydrological characteristics in the watershed are severely hampered due to increased human activities. It is, therefore, recommended that afforestation in the degraded areas be undertaken to restore the watershed health which could improve on hillslope hydrology.

**Keywords:** land use-cover changes, watershed health, river Rwizi

## 1. Introduction

River systems are crucial for the Earth's landscape development [1]. They play significant roles in the provision of water for domestic and industrial purposes as well as a number of resources to communities [2]. Due to their favorable ecological characteristics associated with humid climate and fertile alluvial soils, most river systems have attracted dense populations [3]. The high population densities within river systems have come with increased human activities [4]. The land use dynamics continue to impact on river catchments with negative repercussions on such fragile ecosystems [5]. River catchments are profoundly impacted by land use and cover change driven by anthropogenic activities [6]. The impact of anthropogenic activities on the riverbank morphodynamic is manifested in two ways [7]. First directly, in form of regulation works, artificial cut-offs, gravel and sand exploitation, dam constructions among others. Secondly, it is manifested indirectly in the form of deforestation, social and economic activities in river catchments [5]. Intensive human activities can indirectly impact more dynamic changes of natural processes, which may alter river processes and their subsequent hydro – ecological services [8].

Land Use and Cover Change (LUCC) is a key driver of environmental problems in river valleys and their catchments [9]. Land uses such as forestry, croplands and settlements have transformed most river catchments [10]. Land use and human activities within a watershed can lead to alteration in landscape's hydrological properties [11]. The surface landscape is often fundamentally altered during economic and social development [1]. The quantity, morphology and structure of river systems are usually inadvertently influenced along with land use changes [4]. Estimating historical LUCC trends is essential in assessing the rate at which change occurs; permit exploration of the drivers of that change and related implications [12]. Due to LUCC implications on watershed degradation, impacts like decline in river water quality and quantity often take route [13]. Degraded watersheds have posed serious problems for the environment and people living within river environments [3]. Effective management of watersheds, therefore, requires an understanding of the hydro – ecological resource changes over time and space amidst human activities [14].

The Rwizi river system navigates through Ankole highlands in South Western Uganda providing a source of water and livelihood to people across the region [15]. At least 12 districts including Rakai, Lyantonde, Isingiro, Lwengo, Kiruhura, Mbarara, Bushenyi, Buhweju, Rwampara, Sheema and Rubirizi depend on Rwizi's water for both domestic and commercial use [16]. The Rwizi river system that covers approximately 8,200 km [15] is under serious threat and is on the verge of drying up due to degradation [17]. River Rwizi, which is a lifeline for more than five million people in southwestern Uganda [16] including over 400,000 people in Mbarara City [18], has had up to 80 percent of its water drying up [17]. In recent years, river Rwizi has been branded as a river on the brink of extinction due to climate change, human encroachment and unsustainable resource utilization [15]. Its degradation is not unique but a common challenge to all river systems around the country [19]. Poor land management on the riverbanks and buffer zone is partly responsible for its catchment degradation [20]. The rapid population growth in the region has significantly increased pollution and untreated effluent discharge into the river system [16], while increased demand for water is causing the river to dry up [17]. The destructive agricultural practices, population pressure and unsustainable activities taking place within the river's catchment areas have led to the deterioration of its water quality and quantity [20]. These challenges are now fueling increased socio-economic hardships for local residents that are now manifesting through increased poverty, and continuous water shortages leading to water rationing in the region [15]. These challenges are likely to be exacerbated by the impacts of climate change and continued biodiversity loss [21]. There is, however, particular dearth of information on the implications of LUCC on the ecosystem health of upper Rwizi macro watershed. This, therefore, calls for an urgent need to undertake a spatial –temporal LUCC analysis on the upper Rwizi macro watershed in order to evaluate the major drivers and related implications of this change.

## 2. Methods and materials

### 2.1 Study area

The study was conducted in the upper Rwizi macro watershed within the Ankole highlands of South Western Uganda situated between latitude 0°36′06″S to 0° 48′85″S and longitudes 30°28′30″E to 30°59′65″E (**Figure 1**). River Rwizi originates from Buhweju hills with various tributaries from Ankore highlands and pours its water in Lake Victoria via the drainage systems of Lakes including Mburo, Kachera and Kijanebalola [17]. The upper Rwizi macro watershed was selected for a detailed study owing to increased degradation that is associated with spatiotemporal land use and cover distribution. The drainage network and topography served as baseline information for establishment of the study area and was delineated automatically from a 30 m Digital Elevation Model (DEM) in ArcGIS 10.5. Topographically, the landscape is a dissected plateau, variant of Koki land system characterized by long straight ridges [15]. The topography comprises mainly extensive flat-topped ridges and hills encircling down land arenas (**Figure 2**) typical of Ankole landscape [19]. The ridges have comb-like appearance due to many gullies on the landscape [22]. The geology of the area consists of a sedimentary rock system of the Precambrian age commonly referred to as Karagwe-Ankolean rock system. Quartz–mica and mica schists, shale, phyllite as well as swamp deposits and alluvium dominate this rock system. The soils of the highlands comprise acric ferralsols, dystric regosols, eutric regosols, gleyic arenosols, gleysols, histosols and leptoslos [20].

The study area experiences a bimodal rainfall pattern [16] with two wet and dry seasons [17]. The principal rainfall season is from late August to early December, while the minor one is from late February to mid-May [20, 21]. The average annual rainfall is 1000 mm per annum [23] however, mean annual rainfall can go as low as 750 mm in the eastern and as high as 1520 mm in western parts of the region [17]. There is a marked dry season experienced from June to August as well as January to

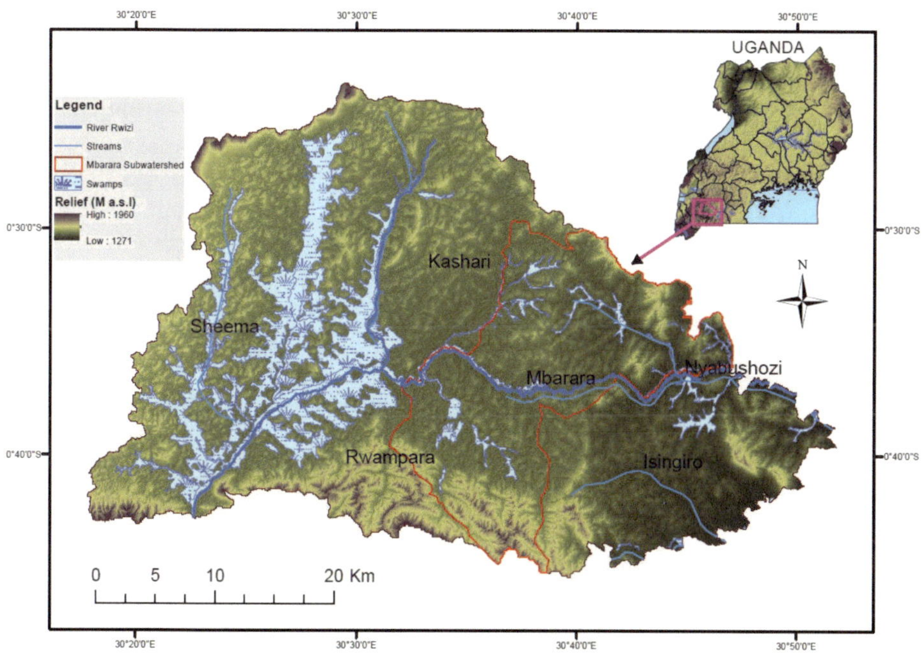


**Figure 1.**
*Location of the study area indicating study sites.*

**Figure 2.**
*Mbarara sub watershed landscape dominated by rolling hills. Photo credit, Ayesiga, November 2020.*

February [24]. More than a century ago, the highlands in the watershed were covered by natural vegetation from the crest to the valley bottom dominated by *Themeda Triandra* and *Hyparrhenia rufa* grass species [21] as well as tree species including *Vernonia ammygdalina*, *Acacia sieberriana*, *Acacia hockii*, *C. edulis* and *Albizia coriaria* [17, 20]. These have however, been colonized by *Londetia kagerensis* on some hillslopes. The valley bottoms are currently dominated by *I. cylindrica* and *Cymbopogan afronadus*/ lemon grass [19]. The vegetation cover is characterized by medium savannah grass with scanty elephant grass (*P. purpureum*). The population density (361/km$^2$) of the region is one of the highest in Uganda [23] with its major source of livelihood being agriculture characterized by arable and livestock farming [16].

### 2.2 Datasets and sources

Remotely sensed data was used to analyze the spatiotemporal land use and land cover distribution for upper Rwizi macro watershed in south western Uganda. The analysis was based on Landsat and Sentinel imagery data sets spanning 1990 to 2020 and 2016 to 2021 respectively. Landsat satellite imagery of 30 m spatial resolution (path 172 and row 060) were downloaded from the open source site- USGS Earth explorer (https://earthexplorer.usgs.gov/). A Key Hole Mark-up Language File (KML) for Rwizi macro-watershed was developed using Google earth and later exported to USGS to delineate the area of interest. Specifications were made about the cloud cover (less than 10%) and dataset requirements. Landsat scenes for 1990 were picked from Landsat 4–5 TM, while scenes for 2000 and 2010 were picked from Landsat 7 ETM$^+$ level 1 archives. Landsat scenes for 2020 were picked from Landsat 8 OLI/TIRS (Operational Land Imager and Thermal Infrared Sensor) satellite. Selected scenes were dropped into the item basket in USGS and downloaded using the Bulk Download Application. The sorted scenes were then prepared in folders in windows files explorer for pre-processing.

#### *2.2.1 Image processing and classification*

Landsat images for 1990, 2000, 2010, and 2020 and Sentinel 2 images for 2016 and 2021 were exported to ArcGIS 10.5, geo-registered to WGS 84 datum, and projected into the Universal Transverse Mercator (UTM) zone 36 N coordinate system. Before performing the final analyses, a series of processes were performed on the downloaded images to improve visualization, processing speed as well as reducing noise. The nature of these processes necessitated swinging between two software packages of ArcGIS 10.5 and QGIS 2.18. The satellite scenes were loaded, visualized and inspected in the two software packages. Gap filling was, therefore, done to seal the existing gaps in the images. This was done in QGIS 3.4.2 using the fill NoData tool under the raster analysis tools provided by the GRASS plugin of open source software. The fill NoData tool which employs data from the gap masks provided by USGS to fill stripped zones on the scenes was done for 2010 image. Satellite imagery scenes downloaded contained the 'black NoData borders' which had to be discarded. This process was carried out in ArcGIS 10.5 utilizing the copy raster tool under the data management toolset. Using this tool, the NoData and background data values were set to zero. The first major pre-processing step performed was to create image composites for ease of visualization and interpretation of selected images. Formation of composites was done by stacking the eight bands of Landsat using the composite bands tool provided under the data management toolset of ArcGIS. Image composites for each path/row combination was done for each of the two scenes for the four selected years. Image composites were created by combining individual bands in a pre-determined pattern. This was done to enhance visualization but most importantly to capture different parts of the electromagnetic spectrum since different objects reflect in different parts of the spectrum. The band combination 432 was selected and used for creating the Red-Green-Blue image composites because it is the best band combination for visualizing and analyzing reflected vegetation [25].

#### *2.2.2 Classification accuracy assessment*

In Ref. [26], overall, producer's and user's accuracies and Kappa coefficient were calculated from the error matrix. The accuracy of the classification was verified by randomly generated reference points using a stratified random algorithm [27]. Following [28, 29], a minimum of 100 random points were generated per class using stratified random sampling approach for accuracy assessment. Field verification for land use and cover classification was conducted between January and August 2020 using draft classified map derived from satellite image for 2020 as a guide. The field verified data were utilized in the maximum likelihood report as an independent data set from which classification accuracy was compared. The land use and cover images for 1990, 2000 and 2010 were validated using Google Earth images taken in November 1990, July 1999 and October 2009 respectively. Higher classification accuracies were obtained for all the downloaded images due to improved sensors. An overall accuracy ranging between 69% (1990) and 85% (2020) was achieved for the Landsat images. The Kappa coefficient accuracies ranged from 0.65 to 0.82 for 1990 and 2020 respectively.

#### *2.2.3 Land-use and cover change detection*

Analysis of land use and cover changes was conducted in three temporal periods including, 1990–2000, 2000–2010 and 2010–2020. The satellite images had to be classified into five thematic classes namely settlements, farmlands, open water, forests and grasslands (Appendix 1). Supervised classification was performed in ArcGIS 10.5 using the maximum likelihood algorithm. The USGS Anderson classification scheme and class definitions were used to assign classes. Post-classification assessments were undertaken using majority filter, boundary cleaning and area calculation to clean the images. Post classification change detection technique was applied to identify the dynamic land use and cover elements from the successive years of satellite data. Post classification was used since the respective imagery were captured using different Landsat sensors and, therefore, with spectral differences. Analysis of land use and cover changes was performed based on automatic comparison of image sub-object hierarchies ("For example, see [30]"). The change detection matrix was computed to determine the proportion of each class which experienced change during the observation period. The major land use and cover change trends were identified from maps generated. The level of persistence was established through a cross-tabulation matrix.

#### *2.2.4 Land use and cover change detection from sentinel images*

Given the spectral characteristics of Landsat imagery, very high resolution images from Sentinel 2 were also used to evaluate the land use and cover changes within particular sections of the river system which could not be detected from Landsat images. Sentinel 2A images of spatial resolution 10 m spanning 2016 to 2021 of MSIL2A_N0214_R078_T36MTE type for 01/05/2016 and 27/01/2021 respectively were acquired from the European Space Agency (ESA) website ((https://sentinels.copernicus.eu/web/sentinel/missions/sentinel-2), through the Sentinels Scientific Data Hub archive. Using the same search specifications as for Landsat, only 2016 and 2021 images were found in the Sentinel 2 hub for Mbarara sub-watershed. Cloud-free Sentinel 2A images covering Mbarara sub-watershed within upper Rwizi macro watershed were selected and downloaded for land use and cover change estimation. Mbarara sub-watershed within upper Rwizi macro-watershed (**Figure 1**) was identified as the most affected section by land use and cover changes over 30 year study period. The European Space Agency's (ESA) satellite constellations Sentinel 2 program was used to combine both high spatial and temporal resolution. The images were subjected to atmospheric correction effects using the Sentinel 2 Correction prototype processing tool in SNAP. The already ortho-rectified Sentinel 2 images were geometrically corrected in Universal Transverse Mercator projection and World Geodetic System (WGS) 84 ellipsoid. The overall accuracy for Sentinel 2 images was 90% for 2016 and 92% for 2021. The Kappa coefficient was 0.87 and 0.90 for 2016 and 2021 respectively. Like for the Landsat images, five thematic classes were generated from the Sentinel images (settlements, farmlands, open water, forests and grasslands) utilizing the USGS Anderson classification scheme.

### 2.3 Field investigations and surveys

Before the interpretation of land use and cover maps, field investigations were undertaken between January and August 2020 to verify the actual land use and cover characteristics. Field verification was also undertaken to confirm whether the created maps lie within corresponding natural boundary. After field verification exercise, some correction was made in land use map before finalizing with the analysis. Additionally, survey data from local people and key informants living within Rwizi macro-watershed was collected to better understand and interpret the LUCC scenarios that emerged from the remote sensing and GIS analysis. During these field investigations conducted with the help of community members and local government technical officials, various land use activities carried out by communities in the watershed were identified and analyzed for their implications on the river ecosystem health. In course of the surveys, the status of riverbanks and buffer zone degradation as well as highly degraded zones within the watershed were ascertained.

## 3. Results

### 3.1 Land use and cover changes between 1990 and 2021

Land use and cover characteristics in the upper Rwizi macro-watershed exhibits both highly spatial and temporal variations. The detected spatial and quantitative changes in land use and land cover between 1990 and 2021 are given in **Figures 3–7**.

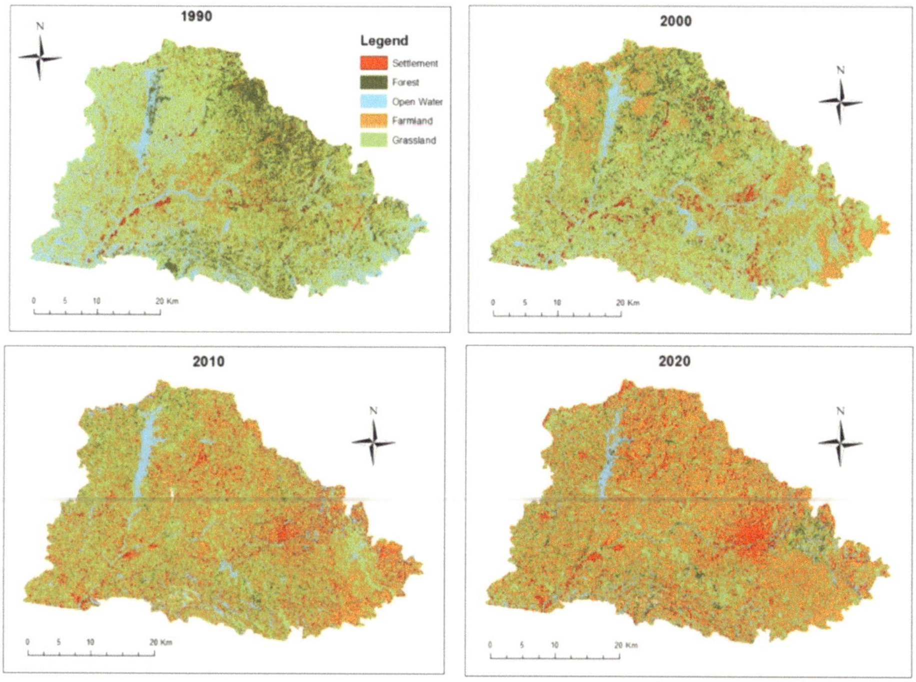


**Figure 3.**
*Land use and land cover distribution between 1990 and 2020 for Upper River Rwizi macro watershed.*

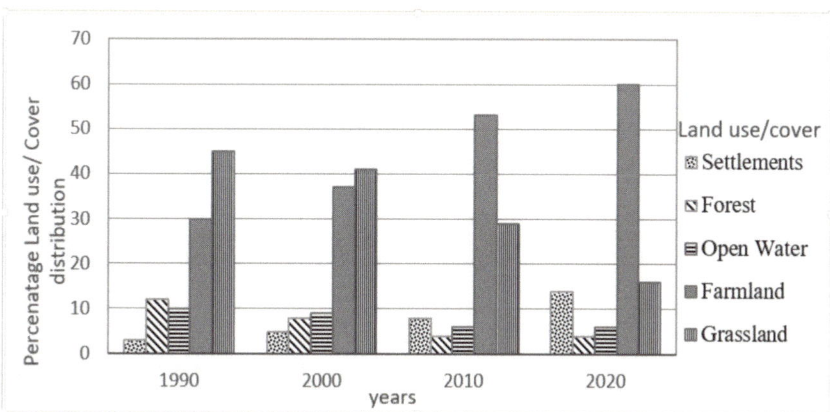


**Figure 4.**
*Land use and cover changes between 1990 and 2020 for upper Rwizi macro watershed.*

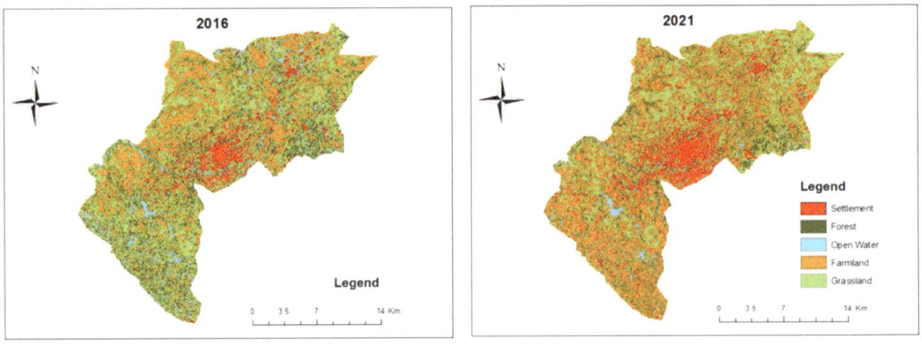


**Figure 5.**
*Land use and cover distribution for Mbarara sub-watershed between 2016 and 2021.*

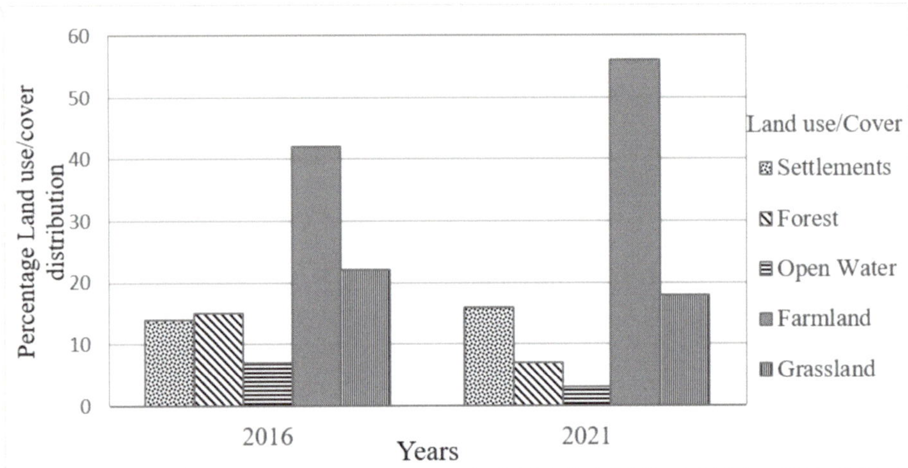


**Figure 6.**
*Land use and cover changes for Mbarara sub-watershed between 2016 and 2021.*

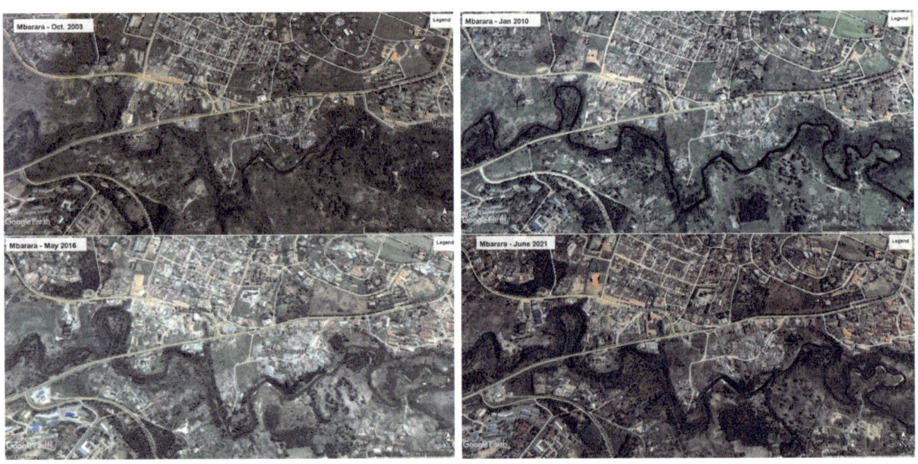


**Figure 7.**
*Land use and land cover distribution between 2003 and 2021 based on Google earth images for a section in Mbarara City.*

There are five land uses and covers detected as shown in the figures. From **Figures 3** and **4**, it is vividly noticeable that whereas, a decrease in forestland and grasslands is noticeable from 1990, farmland and settlements have expanded drastically.

**Figure 4** illustrates the percentage land use and cover distribution in the study area from 1990 to 2020. By 1990, grassland as the dominant land use and cover type spanned 45% of the total study area coverage followed by farmland at 30%. Forest land, open water and settlements covered 12%, 10% and 3% respectively (**Figure 4**). Whereas grassland and forest cover have diminished drastically by 64% and 71% respectively between 1990 and 2020, settlements and farmland have increased tremendously by 79% and 50% respectively over the same period (Appendix 2).

The study area has experienced a drastic conversion of forests and grasslands to settlements and farmlands. For example during the 30 year study period, over 419 $km^2$ of grasslands and 126$km^2$ of forest land have been converted to settlements and farmlands (Appendix 2). The reclaimed open water sections along the river valley have been converted to settlements and farmlands covering 63.8 $km^2$ between 1990 and 2020.

Analysis of land use and cover changes within Mbarara sub-watershed between 2016 and 2021 shows drastic changes in their distribution (**Figures 5** and **6**). During the five year period, forests, open water and grassland reduced by 51%, 54% and 26% respectively within Mbarara sub-watershed. Meanwhile, farmland and settlements increased by 25% and 17% respectively during the same period (Appendix 3). By 2021, farmlands and setllements were the dominant land use and cover types in Mbarara sub-watershed covering 56% and 16% respectively. On the other hand, forests and open water were the least land use and cover types within Mbarara sub-watershed occupying 7% and 3% respectively (**Figure 6** and Appendix 3).

Within the Mbarara sub-watershed, 33$km^2$ of forest land and 28$km^2$ of grasslands were converted into settlements and farmlands between 2016 and 2021 (Appendix 3).

The study area has experienced a drastic conversion from natural land covers especially grasslands and forests to human dominated covers including farmlands and settlements over the past 30 years and more (see Google earth images in **Figure** 7).

### 3.2 Prevailing human activities in the study area

The main human activities in the watershed include crop farming, livestock rearing, fish farming, brick making, sand mining, motor vehicle washing and growing of eucalyptus trees. Forests have been cleared to open up land for farming and settlements. Following field investigations along the Rwizi river valley, it was established that there is increased farming and settlement activities in the study area (**Figure 8**). Reports from the local communities revealed that due to the falling crop yields, many farmers have started encroaching on wetlands along the river valley and forested areas in the search for fertile soils.

The observed farming practices in the study area include both arable and livestock farming. Whereas arable farming is dominant in lowlands and middle slope elements, livestock is commonly practised in the upper slope sections. The major crops grown include both perennial and annuals. The perennial crops include bananas (*Musa paradisiaca*), coffee (*Coffea spp*), tea (*E. abyssinica*), sugarcane (*S. officinarum*) and mangoes (*M. indica*) grown for commercial purposes. The annual crops include Irish potatoes (*Solamum tuberosum*), sweet potatoes (*I. batatas*), maize (*Z. mays*), beans (*P. vulgaris*), peas (*P. sativum*), finger millet (*Eleusin coracana*) and vegetables. In most cases, the crops are inter-cropped. For example farmers inter-crop annual crops in banana gardens especially during the wet season. The dominant cropping systems in the watershed is banana/coffee system.

**Figure 8.**
*Farming activities along Rwizi river buffer zone. Photo credit, Ayesiga, March, 2021.*

The study area has experienced increased population growth with a population density of 361/km$^2$ [23]. The high population growth in the upper river Rwizi macro watershed has led to establishment of several settlements of both rural and urban setting (**Figure 9**). In the study area, settlements are dominant in the valley bottoms and lower slope elements (**Figure 9**). This has led to establishment of roads, buildings and installation of various utilities such as power transmission and telecommunication lines (**Figure 9**).

Due to rapid urban development in the region, there has been an increase in demand for construction materials including sand and bricks. In order to meet the demand for construction materials, local communities have embarked on sand mining (**Figure 10**) and brick making (**Figure 11**) along the riverbanks, wetlands and buffer zones.

### 3.3 Implication of changing LULC

Changes in the land use and cover characteristics associated with increasing human activities have adversely affected the watershed properties. As already noted, there has been unprecedented conversion of natural land covers into human induced types. Following field investigations, it was established that many slopes flanking the river valley appear to be intensively cultivated leading to the emergence of many bare surfaces (**Figure 12**). Increased farming activities have affected

**Figure 9.**
*Urban settlements along Rwizi river valley and buffer zone. Photo credit, Ayesiga patience. August 2020.*

**Figure 10.**
*Sand mining close to river Rwizi near Mbarara City and Rwizi's buffer zone in Nyakayojo. Photo credit, Nseka, December 2020.*

**Figure 11.**
*Brick making site near Ruti in Mbarara City and Nyaihanga in Rwampara Distrct. Photo credit, Ayesiga, February 2021.*

hillslope hydrological properties. This has resulted into increased runoff due to reduced infiltration as evidenced by many gullies and other erosion features on the landscape (**Figure 12**).

The study established that many urban centres have emerged in the study area over the past 30 years. Settlements dominated by urban sprouting have attracted the development of socio-economic infrastructure (**Figure 13**). The construction of socio-economic infrastructure has created paved surfaces that inhibit infiltration thereby increasing surface runoff, erosion and flooding of the lowlands within the watershed.

Due to increasing urbanization, illegal land acquisitions along the river's buffer zone have occurred in the watershed. Several houses as well as crop gardens have illegally been established along the buffer zone within the upper river Rwizi macro watershed. An interaction with communities revealed that illegal land acquisitions along river Rwizi buffer zone has resulted in the reduction of its width and depth (**Figure 13**). Following field investigation and interaction with local communities, it was established that community members no longer need bridges to cross river Rwizi along several of its sections (**Figure 14**). The act of walking in this river is an indicator of low flow velocity and reduced water quantity.

During field surveys, it was established that the river channel is a few metres away from the gardens, however, due to encroachment on the river's banks, it was

**Figure 12.**
*Intensively cultivated slopes which appear to be degraded. Photo credit: Nseka, July 2020.*

**Figure 13.**
*Socio-economic infrastructure sprouting along the Rwizi river valley. Photo credit, Nseka Denis, September 2020.*

hard to realize that there was water flowing. Following an interaction with communities, it was indicated that what used to be papyrus-filled riverbanks has now been cleared and instead, there are water channels from the river flowing to people's gardens. Farmers revealed that they have no alternative to farming other than Rwizi riverbanks, buffer zones and wetlands. Communities living within the upper Rwizi macro watershed confirmed that the water levels in the river system has declined over the years. They, therefore, live in fear that the river will be no more in the

**Figure 14.**
*Shallow section of the river channel where local communities cross without bridges. Photo credit, Ayesiga, September 2020.*

coming years if nothing is done to save it. During interactions with local communities, it was further reported that "the formerly wider river is narrowing". Sometimes, during the dry season, since there is hardly any water flowing from the hills, the water levels reduce into a small stream and, if one is not keen enough, they would think it is not a major river. Local communities indicated that sand mining and brick making along the riverbanks and its buffer zone have almost changed its course. Farming and sand mining are among the most destructive activities chocking river Rwizi to extinction. Due to increased human activities along the river valley, compounded with wetland and swamp destruction, the water levels in the river channel and its tributaries have drastically reduced. This is vividly evidenced by the river's water flow gauging station that is no longer used to monitor flow dynamics (**Figure 15**).

Following field investigations as well as interaction with local communities and key informants, it was established that the water levels of river Rwizi has dwindled by two and a half meters from the highest level where it used to be as indicated by the original gauging pillars at Nyamitanga which were established in 1970

**Figure 15.**
*The river's water flow gauging station near Nyamitanga in Mbarara town has been rendered useless. Photo credit, Ayesiga patience, October 2020.*

(**Figure 15**). Communities living in the watershed revealed that memories of the river with its ample clean water, thick vegetation, wild animals and birds are quickly fading. A close observation of the river channel indicates higher levels of turbidity, with the channel water turning to brown (**Figures 14–16**).

High levels of deposition were also evident in the form of pebbles, silt and clay on the riverbed that has been exposed following low water levels and subsequent reduction in water flow velocity as observed below Rubaya bridge along Buremba road and Ngaromwenda bridge on the Mbarara-Kabale highway (**Figure 16**). The drop in water levels is turning away investors whose project's lifeline depends on good volume of water in the Rwizi river. The drop in water levels is most likely, linked to the degraded hillslopes that no longer have the capacity to recharge the river system.

Due to continuous watershed degradation, many sections of the river channel have been invaded by the waterweed (**Figure 17**). The invasive water hyacinth has congested the river, right from Ndeija sub-county in Rwampara District up to Sango bay in Rakai and the problem has been exacerbated by destructive agricultural practices within the watershed.

The water hyacinth is an indicator of polluted water with many nutrients from the surrounding gardens. Following an interaction with communities, it was revealed that the river's deterioration is due to widespread illegal land grabbing along the fragile riverbanks and buffer zone.

**Figure 16.**
*Dropped water levels along sections of the river Rwizi. Photo credit, Ayesiga patience, December, 2020.*

**Figure 17.**
*Sections of the river invaded and covered by water hyacinth near Mbarara City. Photo credit, Nseka, December 2020.*

## 4. Discussion

### 4.1 Land use and cover changes

The original land cover status for the upper Rwizi macro watershed was generally a natural habitat of diverse flora and fauna experiencing natural hydro-geomorphological processes [21]. The hillslopes were intact, concentrated and promoting the natural movement of water and sediment [15]. The hillslopes flanking the watershed were once well covered by dense vegetation dominated by *Themeda Triandra* and *H. rufa* grass species [21] as well as tree species including *Vernonia ammygdalina*, *Acacia sieberriana*, *Acacia hockii, C. edulis* and *Albizia coriaria* [20] such that surface runoff was limited [17]. The study findings reveal a significant change in the land use and land cover distribution within the upper Rwizi river macro-watershed during the study period. In the study area, it is obvious that trends of land use and cover changes can be observed in the different periods due to evolvement of urban development and farming. Perhaps the most remarkable changes seen in this analysis were in form of settlements and farmlands, which have increased rapidly in the macro watershed. Settlements and farmlands have replaced the pre-existing natural forests, grasslands and open waters. They have directly changed the overall landscape composition of the watershed (**Figures 3–7** and Appendix 2 and 3). The study findings vividly show that there has been a drastic conversion of natural land cover to human manipulated types. It was observed that most of the watershed has undergone massive transformation over the last three decades. These rapid changes could be attributed to the rapid population growth in the watershed over the past few decades [23]. The high population in the region of over 2,574,000 persons [31] and density of 361/ $Km^2$ [16] has put tremendous pressure on the land cover, leading to resource overuse and subsequent degradation [17]. This phenomena has also been responsible for the development of urban settlements (**Figures 9** and **13**). As observed by [6], population and economic growth often lead to spatial expansion of cities. This expansion may occur at the expense of increased risk of flooding, soil loss, riverbank erosion, changes in river courses, sediment load deposition and water pollution [1]. These environmental problems are aggravated by land use change from forestry to agricultural and other development activities such as construction of infrastructural facilities [4]. There has been a significant increase in building density accompanied by large-scale construction of socio-economic infrastructure including roads in the study area [32]. This has resulted into acute shortage of farmlands forcing people to encroach on the sensitive riverbanks and buffer zone [15].

Demographic pressure and the decreasing cultivatable land has significantly contributed to encroachments on both wetlands and the river's buffer zone. Increased demand for farmland and settlement land in the study area has led to uncontrolled clearance of forests and grasslands (**Figures 3–7**). The study findings are in keeping with observations made by [19] who noted that most of the natural land cover in the Rwizi catchment have been destroyed due to increasing land use practices. As can be observed on most of the hillslopes in the watershed, there has been complete depletion of vegetation cover (**Figure 12**). Many hillslopes in the watershed are already bare due to degradation. During the field investigations, it was observed that every available space has been intensively cultivated. Gardens appear like a continuous carpet for kilometers from the valley bottoms to uppermost slope sections (**Figure 12**). The river system has, therefore, been modified to meet the needs of human development in different social development stages. Many people have illegally acquired land along Rwizi riverbanks and buffer zones consequently destroying its wetlands. A number of higher order streams have been invaded, cut off and even buried, forcing most of them to transform into narrower streams. Most wetlands, which work as granaries for water and economically release it during the dry periods, have been destroyed to pave way for settlements and farming activities. As observed by [33], this scenario has partly been due to the existence of informal land acquisition policies and the growing land modification in the region.

### 4.2 Implications of land use and cover changes

The upper river Rwizi macro watershed was under a natural hydrological balance until about 100 years ago, when the dense forests, shrubs, thickets and tall grass began to be depleted [24]. As indicated by [34], the reduction in forest cover greatly affects the landscape's ability to intercept rainfall which greatly affects hillslope hydrological characteristics. Land use and cover changes affect various hillslope hydrological processes including interception, infiltration and evaporation, thereby influencing runoff generation [35]. It has been reported elsewhere that land use and cover changes often lead to natural ecosystem deterioration and landscape degradation [33, 36]. The highlands of the upper river Rwizi macro watershed have been severely degraded as evidenced by bare surfaces in many sections (**Figures** 7 and **12**). This has affected rainfall partitioning thus exposing the hillslopes to severe runoff processes [15]. Vegetation cover degradation which is one of the most important parameters in water and soil conservation [36] has resulted into increased runoff coefficients and soil erosion in the upper Rwizi macro watershed [15].

In Ref. [5] reported a change in the occurrence of the highest runoff and erosion processes incidences in the highlands of South Western Uganda (location of study area) during the year. During the 1960s and earlier when the status of vegetation cover on the valley side slopes was still modest to abundant, the main runoff and erosion risk used to be during the period of March/April [15]. The pattern has, however, changed due to encroachment and the decimation of vegetation cover on valley side and upper slopes in the highlands [19]. The period of worst erosion problem has changed to September/October. This is due to heavy runoff from intense storms that are experienced after the aggressive dry season (June to August) that rage down the almost bare slopes [24]. The change in the main runoff and erosion risk is due to bush burning which is dominant during the prolonged dry season. Bush burning is practiced to provide fresh and greener pastures for the animals and prepare the land for cultivation [21]. Bush burning exposes the land surface to severe runoff and erosion during the subsequent September to November (SON) wet season in the region [17]. This explains why runoff and erosion processes are very intensive at the onset of the SON rainy season. Runoff and erosion processes, however, reduce as the rainfall season proceeds. This is associated with new vegetation that emerges due to increased moisture [15]. The emerging vegetation encourages infiltration while reducing runoff processes [37].

The change in land use and land cover has had significant negative impacts on the soil and river's water quality within the watershed. This is due to the application of excessive amounts of chemical fertilizers and pesticides, which inevitably runoff into the river system. It was reported during community interactions that whenever it rains, water runs on bare slopes, carrying topsoil, pollutants and silt with it. Heavy metals, fertilizers and herbicide have contaminated the waters, thus damaging ecosystems dependent on Rwizi river waters [24]. The silt is affecting the quality of the water in the river system [17]. The expansion of settlements threaten the river ecosystem health, particularly increased sewage disposal and solid waste dumping in rivers. People have built in the river's catchment, thus interfering with the water flow. Increasing urbanization taking place is affecting the hydrological characteristics of a watershed by reducing infiltration of rainwater into the ground and increasing the volume and speed of surface runoff. The replacement of forest cover with paved surfaces or other land use types increases water yield due to reduction in water losses resulting from soil compaction [38]. This increases stream discharge, which is an important element in fluvial processes of erosion and sediment transportation [2]. The processes of erosion and sediment transport result from an interrelated set of natural, human and hydrologic factors within a watershed [10]. The increasing artificial surfaces could be the potential source of pollution within the watershed. Several studies have also shown significant impacts of land use and land cover changes as a contributing source for water pollution [3, 5, 9]. Conversion of the existing land cover increases exposure to other problems in a watershed [7]. This is further confirmed by [11] who maintains that clearance of forest cover accelerates geomorphic processes leading to high sediment yield, rapid channel degradation and mass wasting. In Ref. [13], confirms that destruction of wetlands that can hold water and allow it to flow naturally leads to fast-flowing water in the river system. Due to encroachment within the river Rwizi watershed, the wetland and forest cover along the river valley can no longer hold water and release it in the dry periods. The study established that rainwater flows out of the river system very fast. This situation could contribute to other possible environmental problems such as microclimate change, increased run-off, pollution and degradation of ecological values as well as services within the watershed. Such scenarios could translate into increased hardships for local residents that manifest through decreasing access to clean water, flooding, erosion and intensified salination.

## 5. Conclusion

The upper river Rwizi macro watershed has experienced a drastic decimation of natural land cover categories especially forests and grasslands to human manipulated types dominated by settlements and farming due to increased population amidst informal land acquisition practices. The dominant human activities in the watershed include crop farming, livestock rearing, fish farming, construction, brick making and sand mining. Changes in the land use and cover characteristics associated with increasing human activities have adversely affected the upper Rwizi macro watershed properties. The watershed ecosystem health is worrying due land cover degradation. Due to increased human activities along the river valley compounded with wetland and swamp destruction, the water levels in the river channel and its tributaries have drastically reduced. The drop in the water levels is also due to the hillslopes inability to recharge the river system. The hillslope hydrological characteristics in the watershed are severely hampered. It is, therefore, recommended that afforestation be undertaken in the degraded areas of the watershed. The responsible agencies should also put more efforts into proper implementation and monitoring of already established laws and regulations like the National Environment Regulations, 2000 for wetlands, riverbanks and lake shores management. This will help to restore vegetation cover which could improve on hillslope hydrological balance. Restoration of the vegetation cover will counteract runoff and erosion processes which are on the increase. There is also an urgent need to provide alternative livelihood to river dependent communities in the watershed. With Mbarara's transformation into a city, there is need to reduce the scale of illegal settlements and development activities encroaching into the adjacent river valley wetlands. This study, therefore, recommends a change in land acquisition policies particularly in wetland and forested areas by revising the overlapping mandates of national agencies and district land boards that issue land titles in these fragile sensitive ecosystems. This should be coupled with a strict adherence of watershed developers to Environmental Impact Assessment standards.

## Acknowledgements

The authors gratefully acknowledge the research grant from the Government of the Republic of Uganda (2020/2021) through Makerere University Research and Innovation fund which funded data collection activities and analysis. The authors also gratefully thank the community members within the upper Rwizi macro watershed and the local government technical personnel who provided data and participated in the field surveys and discussions.

## Conflict of interest

No conflict of interest by the authors.

## List of appendices

### Appendix 1: Land use and cover description

| Class | Land-use/cover classes | Description |
|---|---|---|
| 1 | Settlements | Built-up areas, residential, commercials, rural & urban non-residential, roads and other structures |
| 2 | Forests | Tropical, deciduous, coniferous, and plantation forests |
| 3 | Open water | Seasonal and permanent wetlands, swamps, bog, streams and rivers |
| 4 | Farmlands | Cultivated gardens, fallow lands, plantations |
| 5 | Grasslands | Short and tall grasses, thickets, shrubs |

### Appendix 2: Land Cover changes based on Landsat imagery (1990–2020)

| Class | Land Use/ cover class | 1990 | | 2000 | | 2010 | | 2020 | | Overall change | |
|---|---|---|---|---|---|---|---|---|---|---|---|
| | | Sq.km | % age | Sq.km | % age | Sq.km | % age | Sq.km | % age | Sq. km | % age |
| 1 | Settlements | 43.33 | 3 | 65.2 | 5 | 123.11 | 8 | 210.15 | 14 | 166.82 | 79.3 |
| 2 | Forest | 177.4 | 12 | 130.3 | 8 | 51.16 | 4 | 61.71 | 4 | 126.24 | 71.2 |
| 3 | Open Water | 144.5 | 10 | 126.31 | 9 | 93.12 | 6 | 80.72 | 6 | 63.8 | 44.1 |
| 4 | Farmland | 443.4 | 30 | 534.82 | 37 | 780.72 | 53 | 875.67 | 60 | 432.3 | 50 |
| 5 | Grassland | 658.04 | 45 | 610.02 | 41 | 418.46 | 29 | 238.32 | 16 | 419.72 | 64 |
| Total | | 1466.57 | 100 | 1466.57 | 100 | 1466.57 | 100 | 1466.57 | 100 | | |

## Appendix 3: Land cover changes based on Sentinel imagery data (2016 to 2021)

| Class | Land Use/cover class | 2016 | | 2021 | | Overall change over 5 years | |
|---|---|---|---|---|---|---|---|
| | | sq,km | %age | sq.km | %age | Sq.km | %age |
| 1 | Settlements | 59.79 | 14 | 72.11 | 16 | 12.32 | 17 |
| 2 | Forest | 65.64 | 15 | 32.45 | 7 | 33.2 | 51 |
| 3 | Open Water | 24.67 | 7 | 11.36 | 3 | 13.31 | 54 |
| 4 | Farmland | 184.23 | 42 | 245.43 | 56 | 61.2 | 25 |
| 5 | Grassland | 106.44 | 22 | 78.42 | 18 | 28.02 | 26 |
| Total | | 440.77 | 100 | 440.77 | 100 | | |

## Author details

Denis Nseka[1]*, Hosea Opedes[1], Frank Mugagga[1], Patience Ayesiga[2], Henry Semakula[1], Hannington Wasswa[1] and Daniel Ologe[1]

1 Department of Geography, Geo-Informatics and Climatic Sciences, School of Forestry, Environmental and Geographical Sciences, Makerere University, Kampala, Uganda

2 Department of Geography, Faculty of Education, Bishop Stuart University, Mbarara, Uganda

*Address all correspondence to: chiengologe19@gmail.com

## References

[1] Deng X.J, Xu Y.P, Han L.F, Song S, Liu Y, Li G, Wang Y.F. 2015. Impacts of Urbanization on River Systems in the Taihu Region, China. Water 2015, 7, 1340–1358.

[2] Han L.F, Xu Y.P, Lei C.C, Liu Y, Deng X.J, Hu C.S, Xu G.L. 2016. Degrading river network due to urbanization in Yangtze River Delta. J. Geogr. Sci. 2016, 26, 694–706.

[3] Lei Wu, Youpeng Xu, Jia Yuan, Yu Xu, Qiang Wang, Xing Xu & Haiyan Wen. 2018. Impacts of Land Use Change on River Systems for a River Network Plain

[4] Uddin K, Abdul Matin M, Maharjan S. Assessment of Land Cover Change and Its Impact on Changes in Soil Erosion Risk in Nepal. Sustainability. 2018; **10**(12):4715. https://doi.org/10.3390/su10124715.

[5] Wu L, Xu Y.P, Xu Y, Yuan J, Xiang J, Xu X, Xu Y. 2018. Impact of rapid urbanization on river system in river network plain. Acta Geogr. Sin. 2018, 73, 104–114.

[6] Deng X.J, Xu Y.P, Han L.F, Li G, Wang Y.F, Xiang J, Xu G.L. 2016. Spatial-temporal changes of river systems in Jiaxing under the background of urbanization. Acta Geogr. Sin. 2016, 71, 75–85.

[7] Abdulai A. Tahiru, Dzigbodi A. Doke, Bernard N. Baatuuwie. 2020. Effect of land use and land cover changes on water quality in the Nawuni Catchment of the White Volta Basin, Northern Region, Ghana Applied Water Science 10:198 https://doi.org/10.1007/s13201-020-01272-6.

[8] Hai D, Umeda S. & Yuhi M. 2019. Morphological Changes of the Lower Tedori River, Japan, over 50 Years. Water, 11, 1-16.

[9] Ahmed A & Dinye R. 2012. Impact of land use activities on Subin and Aboabo Rivers in Kumasi Metropolis. J Water Resour Environ Eng 4(7):241–251

[10] Chalise D, Kumar L. 2020. Land use change affects water erosion in the Nepal Himalayas. PLoS ONE 15(4): e0231692. https://doi.org/10.1371/journal.pone.0231692

[11] Chalise D, Kumar L, Kristiansen P. 2019. Land Degradation by Soil Erosion in Nepal: A Review. Soil Systems. 3(1): 12. https://doi.org/10.3390/soilsystems3010012.

[12] Wan Yusryza, Wan Ibrahim & Ahmad Nazri Muhamad Ludin. 2016. Spatiotemporal land use and land cover change in Major river basins in comprehensive development AREA, Journal of the Malaysian Institute of Planners SPECIAL ISSUE IV (2016). 225 – 242

[13] Adhikari S, Shrestha S.M, Singh R, Upadhaya S, Stapp J.R. 2016. Land Use Change at Sub-Watershed Level. Hydrol Current Res 7: 256. doi: 10.4172/2157-7587.1000256

[14] Julian J.P, Wilgruber N.A, de Beurs K.M, Mayer P.M, Jawarneh R.N. 2015. Long-term impacts of land cover changes on stream channel loss. Sci. Total Environ. 2015, 537, 399–410.

[15] NEMA (National Environment Management Authority). 2018. State of environment report for Uganda for 2017/18. National Environment Management Authority, Kampala, Uganda. Available at: http://www.nemaug.org [Accessed: 2020-11-17]

[16] UBOS (Uganda Bureau of Statistics). 2020. Statistical abstracts 2020. Ministry of Finance, Planning and Economic Development, Uganda.

Available at: http://www.ubos.org [Accessed: 2021-03-10]

[17] NEMA (National Environment Management Authority). 2020. State of environment report for Uganda for 2019/20. National Environment Management Authority, Kampala, Uganda. Available at: http://www.nemaug.org [Accessed: 2021-04-08]

[18] UBOS (Uganda Bureau of Statistics). 2019. Statistical abstracts 2019. Ministry of Finance, Planning and Economic Development, Uganda. Available at: http://www.ubos.org [Accessed: 2020-10-13]

[19] Wanyama J. 2012. Effect of Land-Use/Cover Change on Land Degradation in the Lake Victoria Basin: Case of Upper Rwizi Catchment, South Western Uganda [PhD disseration], Katholieke Universiteit Leuven.

[20] Bamutaze Y, Wanyama J, Diekrugger, B, Meadows, M., and Opedes H.2017. Dynamics of surface runoff and soil loss from a toposequence under varied land use practices in Rwizi catchment, Lake Victoria Basin. Journal of soil and water conservation, 72 (5): 480-492.

[21] NEMA (National Environment Management Authority). 2016. State of environment report for Uganda for 2015/16. National Environment Management Authority, Kampala, Uganda. Available at: http://www.nemaug.org [Accessed: 2020-07-12]

[22] NEMA (National Environment Management Authority). 2016. State of environment report for Uganda for 2015/16. National Environment Management Authority, Kampala, Uganda. Available at: http://www.nemaug.org [Accessed: 2020-12-10]

[23] UBOS (Uganda Bureau of Statistics). 2018. Statistical abstracts 2018. Ministry of Finance, Planning and Economic Development, Uganda. Available at: http://www.ubos.org [Accessed: 2020-10-20]

[24] NEMA (National Environment Management Authority). 2017. State of environment report for Uganda for 2016/17. National Environment Management Authority, Kampala, Uganda. Available at: http://www.nemaug.org [Accessed: 2020-10-17]

[25] Horning N. 2004. Justification for using photo interpretation methods to interprete Satellite imagery: Version 1.0 American Museaum of Natural History, Center for Biodiversity and Conservation. Available at: http://biodiversityinformatics.amnh.org [Accessed: 2020-12-21]

[26] James, B.C. and Randolph, H.W. 2011. Introduction to remote sensing. The Guilford Press, New York, U.S.A, 335-375.

[27] Jensen J.R. 2005. Introductory Digital Image Processing: A Remote Sensing Perspective. Pearson Education, Inc., New Jersey, U.S.A., 107-312.

[28] Skirvin S.M, Kepner W.G, Marsh S. E, Drake S.E, Maingi J.K, Edmonds C.M, Watts C.J & Williams D.R. 2004. Assessing the accuracy of satellite – derived land – Cover classification using historical aerial photography, Digital orthophoto quadrangles and air borne video data. In R. Lunetta and J.C. Lyon (Eds.), Remote Sensing and GIS Accuracy Assessment, CRC Press, Bocca Raton, Florida, pp. 115 –131.

[29] Congalton R.G & Green K. 2009. Assessing the accuracy of remotely sensed data: Principles and practices (2nd Edition), Taylor and Francis Group, LLC, New York.

[30] Yang H, Adler R, Huffman G. 2007. Use of satellite remote sensing in the mapping of global landslide susceptibility. Natural Hazards 43(2):

245–256. doi: 10.1007/s11069-006-9104-z

[31] UBOS (Uganda Bureau of Statistics). 2017. Statistical abstracts 2017. Ministry of Finance, Planning and Economic Development, Uganda. Available at: http://www.ubos.org [Accessed: 2020-10-12]

[32] UBOS (Uganda Bureau of Statistics). 2016. Statistical abstracts 2016. Ministry of Finance, Planning and Economic Development, Uganda. Available at: http://www.ubos.org [Accessed: 2020-11-17]

[33] Wasswa H, Kakembo V & Mugagga F. 2019. A spatial and temporal assessment of wetland loss to development projects: the case of the Kampala–Mukono Corridor wetlands in Uganda. International Journal of Environmental Studies. 76(2):195–212. Routledge, Taylor and Francis Group Publishers. DOI: 10.1080/00207233.2018.1494931

[34] Josephat M. 2018. Deforestation in Uganda: population increase, forests loss and climate change. Petroleum Production and Geoscience- Makerere University, Department of Wildlife and Natural Resources- Uganda Wildlife. Research & Training Institute, Uganda. ISSN: 2529-8046

[35] Promper C, Puissant A, Malet J.P & Glade T. 2014. Analysis of land cover changes in the past and the future as contribution to landslide risk scenarios, Applied Geography 11–19

[36] Mango L.M, Melesse A.M, McClain M.E, Gann D, Setegn S.G. 2011. Land use and climate change impacts on the hydrology of the upper Mara River Basin, Kenya: results of a modelling study to support better resource management. Hydrology of Earth System. Science. 15 (7), 2245–2258.

[37] Bagoora F.D.K. 1993. An assessment of some causes and effects of soil erosion hazard in Kabale highlands, South Western Uganda, and people's attitude towards conservation. In Abdellatif (Ed) Resource Use and Conservation, Vol 8: Faculty of Social Sciences; Mohammed V. University, Rabat Morocco. Mountain Research and Development.

[38] Mao D & Cherkauer KA. 2009. Impacts of land-use change on hydrologic responses in the Great Lakes region. Journal of Hydrology. 374 (1–2), 71–82.

Chapter 4

# Assessment of Water Quality with Special Reference to Hydrochemistry: A Case Study of Auranga Estuary, Valsad, Gujarat, India

*Shefali S. Patel and Susmita Sahoo*


## Abstract

An assessment of Water Quality from Auranga estuary (20°63' N and 72°820 E) was carried out from January 2019 to December 2019. The hydro-chemical variables were analyzed for the evaluation of water quality showed fluctuation in the estuarine water seasonally. The water quality index was computed for the evaluation of water quality of Auranga estuary; to know the pollution level of water body Index for Pollution was also computed. The water quality index (WQI) was 115.97 at downstream and 85.30 at upstream that indicate poor and good water quality respectively. The Pollution Index (PI) ranges from 1.41 (Downstream) to 0.78 (Upstream) which indicate that the water is medium polluted and slightly polluted respectively. Seasonal assessment showed the discrete water quality index and pollution index based on three different seasons; during winter season WQI was 143.30 and 108.05 and PI was 1.41 and 0.97 at downstream and upstream respectively, during summer WQI was 126.73 and 106.95 and PI was 1.18 and 0.94 at downstream and upstream sites respectively and during monsoon WQI was 97.67 and 88.11 and PI was 0.88 and 0.78 at downstream and upstream sites respectively. Univariate statistical technique is attempted to explain the correlations between the variables.


**Keywords:** Auranga estuary, Hydrochemical status, water quality index, pollution index

## 1. Introduction

Marine and freshwater both ecosystems are incorporated to the estuarine ecosystem and make it a functional water body [1]. The productivity and sustainability of coastal, marine and estuarine ecosystem largely depends on the coastal water quality [2]. Estuaries support important biogeochemical processes that are central to the planet's functioning, e.g. nutrient cycling. As estuaries concentrate waters from very large land surfaces into relatively small water bodies; the biogeochemical processes and trophic interactions within estuaries can play an important role in the management of water quality problems [3]. The surrounding civilization of mankind and their productivity are significantly affected by the water quality. Although, in the developing countries the rivers have faced consequential problems due to the rigorous anthropogenic activities along the water bodies [4]. The seagoing rivers received pollutant components from the transition zone of sea water and fresh water as well as marine pollutants due to the sea water intrusion. There should be more consciousness paid to the spatio-temporal assessment of water quality because of their extensive economic expense and several ecological roles. The most effective and common practice for evaluation of environmental problems is long-term assessment of water quality since the spatio-temopral fluctuations of hydrochemical parameters and biological variables can be conferred clearly and can help in future for research on evaluation of polluation status [4]. The hydr-chemical status of estuaries varies both temporally and spatially and the quality of water is usually described based on its physical, chemical and biological factors and in accordance to the distance from the coastline area as well as tidal phase various responses showed by the water quality of water bodies. The dissolved elemental loadings in the estuaries vary spatially (based on distance from mouth) and temporally (diurnal, seasonal and inter-annum) as well as with depth and laterally, hence the estuary is more fluctuated than sea waters.

The water quality index (WQI) is a spatial trend for determining the depletion of water quality because of the locus of crucial pollutant resources. Water quality evaluation is required for the investigation of water quality, water pollution and their acquaintance of main pollution effective causes, as well as the recognization of polluted risky regions through that polluted surrounding water bodies as well as scrutiny of the diversity of living components [5]. From diverse kinds of pollutions, aquatic environmental pollution has a vital threat to mankind health, also becomes the most remarkable issue for the sustainable development [6]. To determine the water quality of estuaries several factors perform significant role including the quanity and quality of fresh water as well as marine water, biological systems and water circulation and movement. The alternations in these systems can be natural or affected by civilization and artificial activities as well as their catchments [7]. Varol et al. pointed out in 2012 that in times ahead the freshwater is becoming a scanty source with monitoring of water quality and it is a very serious matter in the current decades [8]. The estuarine environment is a very vital for the procreation and natural susceptibility to diequipoise of environment of many marine organisms; that's why there is a need for special attention to estuaries [9]. The rainy and post-rainy seasons significantly influence the freshwater inflow to the estuarine area. The hydrochemical qualities of water body varied according to varying freshwater flow and such variations depend on several ecological reactions like fluctuations in composition of species, growth of phytoplankton blooms and reduction of oxygen concentrations. The fluctuation in water quality is a continuous process [10]. Tropical estuaries have more pronounced nutrient dynamics as well as tidal variation in comparison to temperate estuaries [11]. For sustainable utilization and conservation of water resources it is necessassery to identify the water pollutants and spatio-temporal assessment of water quality [10]. The water temperature and discharge found affiliated to the climatic constituents include hydrochemical and biological systems as well as flow regulation that is closely linked to the seasonal variation of water quality [12].

The rapid growth of urban areas and industries has an excess demand of water and that is why limited water resources are under tremendous pressure. An urban land scenery established due to the higher developed industrialization in India and it gives rise to problems on water pollution dangerous to all living beings. The water resources are getting polluted due to the untreated effluent discharges and industrial dumping off in to the water pits and water channels [13]. Mankind activities such as garbage dumpings, utilization of agricultural fertilizers and chemicals with its rapid utilization of water resources affect the water quality [6]. A large number of components and corresponding evaluation factors are necessary for the assessment of water quality at the basin scale, as well as a geographical distribution of pollution levels based on every component and evaluation factor. Understanding the temporal trends and spatial distributions of water quality are essential for maintaining the health of aquatic ecosystems and ensuring the safety of water [14]. Decline in water quality is mainly due to the increased concentration of various pollutants such as oils, heavy metals, nutrients and organic compounds [15]. Deterioration of water quality also occurred through the efflux of suspended solids, such as erosion from river banks, sediment and silt, agricultural discharges, wash-off from logging fields and construction sites, which are affecting the various bodies of water, thereby aquaculture respiration becomes impaired, primary productivity and depth of water bodies become reduced, and aquatic ecosystem become suffocated.

There are a large amount of complicated phenomenon and multiple-factor in comprehensive water quality estimation, and many more fuzzy concepts and phenomena are entailed in an evaluation. The water quality index (WQI) is a numeric expression which is used for the water quality assessment of a selected water body; thereby it is possible to easily understand the condition of a water body by managers from many countries [16]. At first Horton evolved WQI in 1965 in the United states by choosing the 10 most commonly used factors for the assessment of water quality such as Chloride loadings, Alkalinity amount, pH and Dissolved oxygen concentration, etc. and it has been broadly used and undertaken in Asian, European and African countries [17].

Nowadays, the literature on water quality from inland rivers and coastaline areas can be obtained easily, however there are very few reports on information of sea Voyage Rivers because of their complex ecosystem and diverse pollution resources. Literature on some of the spatio-temporal assessment of estuarine ecosystems includes Motru river estuary, Romania [16]; Estuarine systems from South Africa [7]; Estuary of Duliujian river, China [4]; Scheldt estuary, Belgium and Netherlands [3]; Zhangweinan river basin, china [10]; Ying river basin, china [12]; Tapi etsury, west coast of India [11]; Narmada estuary, Gujarat, India [18]; Bay of Bengal, India [19].

One of the fundamental steps in establishing such an integrated water management approach is the development of an integrated monitoring programme. This study explicitly considers the spatio-temporal dependence model for overcoming spatio-temporal variability. The relationship of water quality parameters observations vary with time and space. The water quality of the Auranga estuary from western India was assessed according to spatial and temporal fluctuations and was the main aim of this case study. A WQI considers the spatio-temporal coefficient can sufficiently explain the spatial and temporal variations in the quality of water from a target area. This research aims to reveal the water quality status based on WQI. The objective of this study is to find out the main pollutants in different seasons and estuarine reaches by multivariate statistical methods, which are expected to help managers to understand the water body system along the estuary.

## 2. Materials and methods

### 2.1 Study area

Western Indian state Gujarat has 33 districts and Valsad district is one of them. There is the longest coastline about 1600 km of India occupied by Gujarat; of that 73 km of coastline is occupied by the Valsad district (**Figure 1(a)**). Auranga river originates from near Bhervi village and flows through Valsad taluka as well as city and ends into the Arabian Sea (**Figure 1(b)**). It flows across Kosamba, Bhadeli, Lilapore, Vejalpore, Gundlav, Abrama, Jujwa, Ghadoi, Kalwada, Nandhai, and Bhervi villages. The Auranga river has 97 km in length with 699 sq./km catchment area. AN estuary of Auranga is located under 20°63' N Latitude and 72°82′ E Longitude. Tithla beach is located near the Auranga estuary. This river is very important for the socio-economic life in the southern Gujarat.

The water resource of Auranga hydrographical complex is providing the drinking water supply as well as the exploitation of water for industrialization as well for domestic purposes, which can influence the hydro-morphological characteristics of tributary, changing the conditions of the liberated natural water regime on their courses.

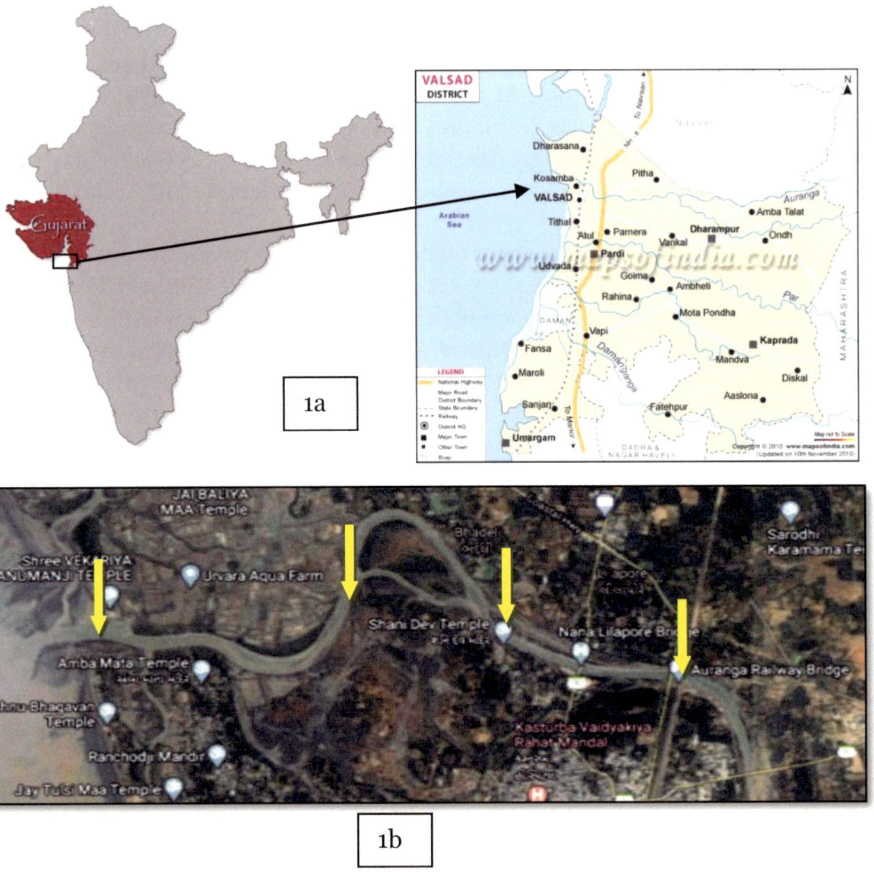


**Figure 1.**
*(1a) – Geographical locations of Gujarat in India and Valsad District in Gujarat; (1b) – Estuarine area and sampling sites from Auranga estuary.*

### 2.2 Sample collection

The sampling activity was conducted from four sites (**Figure 1(b)**) of the solely flowing Auranga estuary based on bi-monthly intervals from January to December 2019 which include all three seasons; winter, summer, and monsoon. A total of 24 samples were analyzed that includes physicochemical parameters including heavy metals. Three samples were collected from each sampling site and then made into composite samples. The sampling sites were selected based on resources available for sampling, experimental sampling, and checked water quality parameters. The water samples were collected by using 1-liter clean polythene bottles and stored in an ice box at 4° C temperature and then transported to the laboratory as soon as possible for physicochemical parameters analysis. The estuarine water samples were fixed in 300 ml BOD bottles for the immediate estimation of dissolved oxygen and measurement of biochemical oxygen demand after 5 days of incubation at 20° C in an incubator.

### 2.3 Water quality assessment

Hydro-chemical parameters were analyzed in 7 to 10 days immediately after the water samples were collected, the analysis methodology strictly follows the protocols described in manuals [20, 21]. Water temperature was determined by Thermometer, Electrical Conductivity was determined through the digital conductivity meter and pH was measured by digital pH meter on-site during sampling. DO was analyzed on-site and BOD was analyzed after 5 days incubation period through the Winkler method. Other physicochemical parameters analysis was carried out in the NVPAS laboratory. Turbidity was determined with the Turbidity meter. Alkalinity, Salinity, Free $CO_2$, Chloride, Total Hardness, and Ca Hardness were analyzed through the titration method. COD, Ammonium, and Color were analyzed in SICART. Nitrate, Inorganic Phosphate, and Silicate were determined with the help of the Spectrophotometric method. The gravimetric method was applied for the assessment of TSS, TS, and TDS. A flame photometer was used to assess the Sodium and Potassium concentrations. The Turbidimetric method was implemented for Sulfate assessment. Heavy metals were analyzed through the ICP-OES in SICART.

The Water Quality Index (WQI) is mostly applied for the evaluation of the water quality of a particular water body. The WQI can be divided into five groups, each group with a different quality state and with a different usage domain [22]. Here a mathematical equation incorporates several quantitative variables that give the scale in numbers of the quality of water bodies.

### 2.4 Computation of WQI

The WQI is computed through the following three steps [23].

**First step** – Based on the concerned significance for the overall quality of water, assigning of weight (wi) to the selected water quality variables (e.g., Chloride, pH, Temperature, TDS, Phosphate, Nitrate, Iron, Boron ... ) (weight may be from 1 to 5).

**Second step** – Compute the relative weight (Wi) of the selected variables by using the following equation:

$$Wi = \frac{wi}{\sum wi} \tag{1}$$

(**i = 1 to n;** n = total number of selected variables)
Wi = Relative weight,
Wi = Assigning weight of each variable and 'n' is the number of variables.

| Sr. No. | Parameters | Units | Prescribed standards |
|---|---|---|---|
| 1 | Chloride | mg/L | 250.0*** |
| 2 | Boron | mg/L | 0.7**** |
| 3 | Phosphate | mg/L | 0.001–0.01** |
| 4 | Sulfate | mg/L | 250*** |
| 5 | DO | mg/L | 4.0* |
| 6 | pH | | 6.5–8.5* |
| 7 | Temperature | °C | 26–30** |
| 8 | Nitrate | mg/L | 50*** |
| 9 | TDS | mg/L | 1000*** |
| 10 | Color | Hazen | 150** |
| 11 | Alkalinity | mg/L | 115***** |
| 12 | COD | mg/L | 20*** |
| 13 | Iron | mg/L | 0.3*** |

** Water Quality Standards for Coastal Waters Marine Outfalls. SW-II Standard. Central Pollution Control Board, New Delhi.; ** South African Water Quality Guidelines for Coastal Marine Waters, 1996. International Target Values for the Natural Marine Environment, Vol.1, pp. B-1-B-3. and Chap. 4.2. pp. 31.; *** KepMenKes No. 51/MENKES/SK/VII/2004. quality standards of the Environment Decree No. 51 in 2004 on Marine water quality standard for marine biotas.; **** UK Marine Standards.; ***** Canadian Water quality standards for Marine fisheries and aquatic life, Environment Canada, 1987 CEC, 1978, 1980 committee for fisheries, 1993.*

**Table 1.**
*Water quality standards for coastal waters.*

**Third step** – Find out the quality rate (qi) for each selected variable, as below:

$$Qi = \left(\frac{Ci}{Si}\right) \times 100 \quad (2)$$

Qi = Quality rating scale,
Ci = Concentration of each selected variables.
Si = Guideline value/desirable limit as given in Water Quality Standards (**Table 1**).

**Fourth step** - For the computation of WQI, the sub-index (SI) is first determined for each chemical variable, as follows:

$$SIi = Wi \times Qi \quad (3)$$

$$WQI = \sum_{1}^{n} SIi \quad (4)$$

SIi = Sub index of $i^{th}$ variable;
Wi = Relative weight of $i^{th}$ variable;
qi = Quality rating scale of $i^{th}$ variable and 'n' is the number of selected variables (**Table 2**).

## 2.5 Calculation of pollution index

Single-factor pollution index was formulated as [24]:

$$\mathrm{Pi} = \mathrm{Ci}/\mathrm{Si}$$

| CWQI- range | Category-rank |
|---|---|
| <50 | Excellent water |
| 50–100 | Good water |
| 100–200 | Poor water |
| 200–300 | Very poor water |
| > 300 | Unsuitable |

**Table 2.**
*Coastal water quality ranking criteria [15].*

| Pollution index (PI) | Pollution category |
|---|---|
| 0.4 | Not-polluted |
| 0.4 ∼ 1.0 | Slightly polluted |
| 1.0 ∼ 2.0 | Medium polluted |
| 2.0∼ 5.0 | Highly polluted |
| >5.0 | Serious polluted |
| *DOI:10.1371/journal.pone.0119130.t002* | |

**Table 3.**
*Standard values of pollution index.*

Where, Pi = the pollution index.

Ci = the measured concentration of variables (pollutants).

Si = Guideline value/desirable limit as given in Water Quality Standards (**Table 3**).

All mathematical and statistical computations were made by PAST 3.0.

## 3. Result and discussion

To protect the environment in general and preserve good water quality in particular, an effective and modern water management system is necessary. For the monitoring of water quality of Auranga estuary the water quality index was computed for the period of a year 2019 from two sampling sites; downstream near Divadandi light house, Kosamba, and upstream near Valsad water treatment plant, Abrama taking into account the maximum annual, the minimum annual, and the mean annual values of 26 following physical and chemical parameters: Color (Hazen), Alkalinity (mg/L), Free Carbon Dioxide (mg/L), Electric Conductivity (μs/cm), Turbidity (NTU), Temperature (°C), pH, Salinity (ppt), Silicate (mg/L), Hardness (mg/L), TDS (mg/L), TSS (mg/L), DO (mg/L), BOD (mg/L), COD (mg/L), Chloride (mg/L), Sulfate (mg/L), Sodium (mg/L), Potassium (mg/L), Nitrogenous compounds such as Ammonium (mg/L) and Nitrate (mg/L), and Phosphate (mg/L) and presented Trace metals (Boron and Iron) (mg/L) with units of measurement adapted according to "International Union of Pure and Applied Chemistry".

Water quality monitoring is not only the scientific description of status of water body, but also revealed the directions for control and management programs of water pollution. From this water quality monitoring program, the appropriate model about water quality can build through the received information. The models of water quality and parameters identification can be determined through the assessment of water quality. This kind of model accurately revealed the status of water body, master resources of pollution, trends of development, and scientific methods to protect the aquatic ecosystem with planning and proper management [25].

In the case of physical parameters, the annual average values of color and turbidity were higher downstream than upstream site. The turbidity is in a dynamic state and is significantly changed during heavy rainfall events. Several factors are affecting the turbidity such as sunbeam, apparent water state as well as the removal of suspended compounds from the terminal water flow and water column. Apart from causing variations in the surface water quality, rainfall can directly increase the level of suspended solids through runoff. Temperature and pH were almost similar at both sites of Auranga estuary based on the annual average value. The meteorological properties fluctuated the temperature. An alkaline feature of this water body was noticed with pH (Average of 7.4, varied from 6.8 to 7.7) is designating that this water body has limited variations of pH values due to the presence of buffering capacity. Similar observation about pH range 7.2–7.6, with high pH in winter was made from the Timis river basin [26]. The effects of pH on phosphate adsorption should be responsible for strong positive correlations of phosphate. In the alkaline state, the ion exchange released phosphorus thereby; metal cations were replaced by $OH^-$ and combined with phosphate consequently leading to more dissolved phosphate contamination [4]. Alkalinity, Chloride, Salinity, and sulfate had high annual average concentrations in downstream than upstream site because of the higher content of seawater than fresh water in the study area. More salinity during the periods of winter and summer was due to the high degree of evaporation of exterior surface water and the adjoining neritic water domination as well as the low-lying wave and tidal activity with reduced freshwater inflow and land drainage. The minimum salinity was during monsoon (**Table 4**) due to the monsoonal rain, flooding, and freshwater input into the study areas. The existence of large amounts of organic materials leads to excessive contamination of chloride in water. A large amount of chloride in water indicated the pollution of animal origin, hence chloride concentration and pollution status have a direct correlation. The poor contamination of chloride during the rainy season (**Table 4**) was possible because not many loadings from the industrial activity while, the higher concentration of it may be due to seawater intrusion coupled with a huge influx of sewage and industrial wastewater [3]. The calcium concentration fluctuates with variables such as land field, precipitation, and dissipation in the coastline waters constituting close to the shore. At the early stretches of estuarine transition, excess calcium amount attributed to their exemption from the interchangeable sites of water body's clays with other cations [3]. The total suspended solids (TSS) increased at the middle of the estuary and further downstream due to the inrush from the upper stretches, wastewater discarding and the usage of traveling boats and fish catching boats for caged fish rearing. The annual average value of electrical conductivity was higher at the downstream site than at the upstream site because of more content of salts. During winter electrical conductivity was recorded higher (**Table 5**) which is attributed to a little mingling of freshwater influx from riverine stretch causing more concentration of ions. Dissolved oxygen (DO), Biological oxygen demand (BOD), and Free carbon dioxide had higher loadings downstream than upstream site, while chemical oxygen demand (COD) had almost similar annual average loading at both sites of Auranga estuary. Decreasing freshwater inflow, land field, anthropogenic sewage, and industrial runoff increases temperature, salinity, and growth of phytoplankton and during decomposition utilization of oxygen through the microbial activity during winter leading to maximum COD and low COD was observed during the monsoon season due to the presence of heavy river run-off, reduction of mixing of domestic and agricultural garbage, land-field drainage into the estuarine ecosystem

| | Color | Alkalinity | TDS | TSS | pH | Temperature | Free $CO_2$ | Salinity | Total hardness | Ca hardness | Turbidity | EC | Chloride | DO | BOD | COD | Phosphate | Nitrate | Ammonium | Sulphate | Sodium | Potassium | Silicate | Boron | Iron |
|---|---|---|---|---|---|---|---|---|---|---|---|---|---|---|---|---|---|---|---|---|---|---|---|---|---|
| Color | 1 | | | | | | | | | | | | | | | | | | | | | | | | |
| Alkalinity | 0.47 | 1 | | | | | | | | | | | | | | | | | | | | | | | |
| TDS | 0.64 | 0.79 | 1 | | | | | | | | | | | | | | | | | | | | | | |
| TSS | 0.33 | 0.90 | 0.51 | 1 | | | | | | | | | | | | | | | | | | | | | |
| pH | −0.87 | 0.32 | 0.56 | 0.17 | 1 | | | | | | | | | | | | | | | | | | | | |
| Temperature | −0.70 | 0.45 | 0.64 | 0.40 | 0.76 | 1 | | | | | | | | | | | | | | | | | | | |
| Free $CO_2$ | 0.81 | 0.88 | 0.85 | 0.77 | 0.64 | 0.71 | 1 | | | | | | | | | | | | | | | | | | |
| Salinity | 0.92 | 0.21 | 0.59 | 0.22 | 0.85 | 0.61 | 0.61 | 1 | | | | | | | | | | | | | | | | | |
| Total hardness | 0.96 | 0.64 | 0.69 | 0.52 | 0.77 | 0.65 | 0.90 | 0.83 | 1 | | | | | | | | | | | | | | | | |
| Ca hardness | 0.93 | 0.52 | 0.61 | 0.35 | 0.76 | 0.84 | 0.78 | 0.85 | 0.95 | 1 | | | | | | | | | | | | | | | |
| Turbidity | 0.89 | 0.30 | 0.64 | 0.40 | 0.91 | 0.57 | 0.62 | 0.22 | 0.80 | 0.87 | 1 | | | | | | | | | | | | | | |
| EC | 0.49 | 0.93 | 0.66 | 0.91 | 0.20 | 0.36 | 0.86 | 0.95 | 0.62 | 0.56 | 0.21 | 1 | | | | | | | | | | | | | |
| Chloride | 0.98 | 0.56 | 0.73 | 0.42 | 0.85 | 0.76 | −0.22 | 0.90 | 0.97 | 0.91 | 0.86 | 0.57 | 1 | | | | | | | | | | | | |
| DO | 0.86 | 0.83 | 0.81 | 0.70 | 0.71 | −0.62 | −0.42 | 0.68 | 0.47 | 0.88 | 0.72 | 0.81 | 0.90 | 1 | | | | | | | | | | | |
| BOD | 0.85 | 0.76 | 0.74 | 0.59 | 0.75 | −0.51 | −0.41 | 0.70 | 0.21 | 0.92 | 0.79 | 0.71 | 0.86 | 0.97 | 1 | | | | | | | | | | |
| COD | 0.47 | 0.03 | 0.34 | 0.12 | 0.83 | −0.63 | 0.26 | 0.55 | 0.31 | 0.32 | 0.68 | 0.22 | 0.46 | 0.30 | 0.37 | 1 | | | | | | | | | |
| Phosphate | 0.93 | −0.57 | 0.73 | 0.47 | −0.33 | −0.78 | 0.88 | 0.83 | 0.94 | −0.83 | 0.74 | 0.63 | 0.97 | 0.86 | 0.77 | 0.30 | 1 | | | | | | | | |
| Nitrate | 0.86 | −0.39 | 0.64 | 0.24 | −0.68 | −0.85 | 0.70 | 0.83 | 0.77 | −0.71 | 0.88 | 0.25 | 0.86 | 0.73 | 0.74 | 0.83 | 0.76 | 1 | | | | | | | |
| Ammonium | 0.39 | 0.25 | 0.01 | 0.38 | 0.73 | −0.26 | −0.01 | 0.51 | 0.71 | −0.32 | 0.64 | −0.41 | 0.30 | 0.12 | 0.27 | 0.88 | 0.10 | 0.66 | 1 | | | | | | |
| Sulfate | 0.52 | 0.99 | 0.82 | 0.86 | 0.34 | 0.46 | 0.70 | 0.48 | 0.69 | 0.60 | 0.37 | 0.94 | 0.61 | 0.87 | 0.80 | 0.13 | 0.62 | 0.40 | −0.25 | 1 | | | | | |
| Sodium | 0.34 | 0.48 | 0.18 | 0.57 | 0.03 | 0.49 | 0.48 | 0.12 | 0.42 | 0.45 | 0.08 | 0.58 | 0.40 | 0.35 | 0.15 | 0.31 | 0.58 | 0.93 | −0.51 | 0.34 | 1 | | | | |

| | Color | Alkalinity | TDS | TSS | pH | Temperature | Free $CO_2$ | Salinity | Total hardness | Ca hardness | Turbidity | EC | Chloride | DO | BOD | COD | Phosphate | Nitrate | Ammonium | Sulphate | Sodium | Potassium | Silicate | Boron | Iron |
|---|---|---|---|---|---|---|---|---|---|---|---|---|---|---|---|---|---|---|---|---|---|---|---|---|---|
| Potassium | 0.45 | 0.16 | 0.41 | 0.31 | 0.07 | 0.17 | 0.34 | 0.55 | 0.47 | 0.68 | 0.15 | 0.32 | 0.47 | 0.32 | 0.22 | 0.29 | 0.60 | 0.66 | −0.34 | 0.22 | 0.43 | 1 | | | |
| Silicate | 0.26 | 0.06 | 0.20 | 0.10 | 0.60 | 0.77 | 0.23 | 0.23 | 0.13 | 0.13 | 0.28 | 0.14 | 0.30 | 0.15 | 0.11 | 0.80 | 0.25 | 0.11 | 0.55 | −0.01 | 0.10 | −0.40 | 1 | | |
| Boron | 0.76 | −0.08 | 0.36 | −0.32 | −0.32 | −0.45 | 0.31 | 0.92 | 0.59 | 0.45 | 0.91 | −0.11 | 0.67 | −0.42 | −0.50 | 0.69 | 0.58 | −0.73 | 0.75 | 0.02 | −0.18 | 0.36 | 0.28 | 1 | |
| Iron | 0.35 | −0.86 | −0.66 | −0.89 | 0.18 | −0.67 | 0.19 | 0.09 | −0.48 | 0.26 | 0.59 | −0.08 | 0.43 | 0.62 | 0.49 | 0.66 | 0.56 | 0.68 | 0.33 | 0.81 | −0.63 | −0.06 | 0.37 | 0.22 | 1 |

**Table 4.**
*Correlation coefficient matrix of Physico-chemical parameters from Auranga estuary.*

| Seasons | Winter | | Summer | | Monsoon | |
|---|---|---|---|---|---|---|
| | Downstream | Upstream | Downstream | Upstream | Downstream | Upstream |
| WOI | 143.30 | 108.05 | 126.73 | 106.95 | 97.67 | 88.11 |
| PI | 1.41 | 0.97 | 1.18 | 0.94 | 0.88 | 0.78 |

**Table 5.**
*Seasonal water quality index and pollution index from both sites of Auranga estuary.*

and reduction of biological activity because of the decreased temperature and salinity. DO plays a vital role as a pivotal indicator of most of the biological, chemical as well as physical systems of the water hence it is considered as the significant variable of water quality [27]. Higher DO concentrations recorded during the winter season (**Table 5**) may be due to the combined effects of higher wind energy and the mixing of heavier rainfall and freshwater. Furthermore, the diversity of aquatic autotrophic components and their ability to produce oxygen may also be another important factor influencing the DO concentration. The periodic freshwater influx and the oxygen utilization by the microbial activities during winter give rise to the higher value of BOD. The maximum dissolved oxygen in February and November was also observed in the Timis river basin which ranked 1st quality, with a declined in August [27]. Silicate, potassium, and sodium had identical annual average contamination at both sites of the estuary. Ammonium, nitrate, and phosphate were similar at both sites of the estuary based on annual average concentration. Incorporation of anthropogenic sewage and agricultural fertilizers and influx from upper reaches lead to excessive inorganic phosphate during winter and summer periods. A high concentration of nitrate observed during the winter season might be due to the resultant freshwater draining, terra drainage, and fertilizers loadings from the nearby agricultural fields and oxidation of ammonia. Most of the nitrate might have been derived from the decomposition of organic wastes [4]. The variables such as temperature and DO showed a relationship with the seasonality, while the total suspended solids, turbidity, and nitrate were correlated with surface runoff caused by rainfall events [28]. In the developing nations, anthropogenic activities are the main sources of heavy pollution in many rivers. Several human activities such as discharges from effluents, agricultural chemicals utilization, over exploitation of water resources influenced the surface water quality. Several factors include total phosphorus, total nitrogen, nitrite and ammonium were the medium to serious scale pollutants in the Honghe river watershed, China [29]. In the case of heavy metal contaminations, Boron had a similar annual average concentration at both sites of Auranga estuary. Iron had higher annual average loading at the upstream site than downstream site during the study period. The water quality status of Zhaoquan River and Wailiao River was good, and Pangxiegou River and Qingshui River represented unsatisfactory water quality status [30]. The varifactors resultant from factor analysis indicated that the variables responsible for water quality changes are mainly associated to discharge and temperature (natural), organic pollution (point source: domestic wastewater) in relatively less polluted areas; organic pollution (point source: domestic wastewater) and nutrients (non-point sources: agriculture and orchard plantations) in medium polluted areas; and organic pollution and nutrients (point sources: domestic wastewater, wastewater treatment plants and industries) in highly polluted areas in the basin [31, 32].

Most of the component loadings were higher in downstream than the upstream site during the study period. In the case of seasonal variations, most of the variables were more in contamination during the winter season but some of the variables were more in concentration during summer (iron) and monsoon (total suspended solids and pH) seasons. The water quality improved a lot in the rainy season due to the dilution of estuarine water through the rainwater. The observation on poor water quality in downstream site than upstream was determined; it was poor grade water quality during winter also found in Yongjiang River, China [33].

Water quality index method can not only give the water quality rank, but also reflect the spatial and temporal variations of water quality condition. The worst water quality attributes to rapid population increase and fast industrial development, which causes the increase of wastewater discharge [34]. The water pollution in India has become a serious issue to economic, social sustainable development, not only because the imbalance between available scant water resources and dense population, but also the inefficient of water resources regulation and management. During the winter season, water quality was poor at both sites of Auranga estuary with more poor (143.30) and less poor (108.05) WQI at the downstream and upstream sites respectively. During the summer season, it was 126.73 for downstream and 106.95 for upstream that also indicates the poor water quality at downstream with medium poor water quality index value and also poor water quality at the upstream site but with less poor water quality index value. The water quality index was 97.67 for the downstream site and 88.11 for the upstream site during the monsoon season which proposed good water quality at both sites of this water body in monsoon season (**Table 5**). Certain WQI scales are the result of changed water quality due tothe wastewater from human activities, industrial establishments, agricultural resources and illegal waste disposal in the basin. The computed average WQI values for the Timok River during the period of 1990–2014 showed significant oscillations in water quality and indicated high level of water pollution with organic and inorganic substances [35]. This study highlighted based on the WQI and PI that the water quality of the Auranga estuary was poor in the winter and summer seasons due to more component contamination and good water quality in the monsoon season due to more freshwater influence because of the heavy rain.

The pollution indexes were 1.41 for the downstream site and 0.97 for the upstream site during winter that proposed medium polluted and slightly polluted water respectively of Auranga estuary during this season. During the summer season pollution indexes were 1.18 and 0.94 for downstream and upstream respectively that also indicate medium polluted and slightly polluted water. During the monsoon season, pollution indexes were 0.88 for the downstream and 0.78 for the upstream site which proposed slightly polluted water at both sites of this water body (**Table 5**). The main water pollution was originated from main two kinds of sources: the natural sources and the anthropogenic sources were associated with urban input, sewage, industrial dumps and surface runoff from agricultural areas [36].

There are mankind actions and demographic properties on one site and urban and partial industrial activities on another site, that is affected the water quality of the Auranga hydrographical basin. In this basin, the surface water resources and groundwater of this region are getting polluted by the main causes such as releasing of untreated wastewater from the industrial areas, domestic garbage, and pollutants from farming activities as well as animal farming activities. The human stress on the surface water within Auranga river catchments is induced by the total number of inhabitants (almost 51,410 people) from villages and the urban inhabitants (almost 1,53,271 people) from Valsad Municipality and Valsad Industrial notified area by census-2011 by the organic loading that they generate through the industrial activities, land use, and animal husbandry, and lastly the hydrographical system improvement through, as an outcome of mankind actions.

Based on the bi-monthly interval assessment, the component loadings were more in November and December. Some of the variables such as total dissolved solids, total suspension solids, chloride, temperature, electrical conductivity;

sodium, and potassium were more concentrated in the months of March and April. The lowest contamination was found during the month of September and October during the study period because of the dilution of water due to the rainy season.

WQI gives important and objective information, and it is worth further promoting water quality inspections. It is a feasible method for evaluating the water quality conditions [30]. The water quality index over the year was 115.97 at downstream and 85.30 at upstream. The water quality status is poor (100–200) at the downstream site and good (50–100) at the upstream site according to the average annual loadings of the analyzed parameters during the study period. The values of the water quality index from these two stations correspond to the poor and good water class, which are influenced by the various variable, by the high values of the chloride, sulfate, alkalinity, iron, nitrate, DO, pH, COD, boron, temperature, phosphate, and TDS from the water of Auranga river estuary, as a result of the agricultural practices, municipal and industrial wastewaters, dumping site, manure from farms, and so on. The pollution index over the year was 1.09 (medium polluted) at downstream and 0.86 (slightly polluted) at upstream of Auranga river estuary. The water quality of the Auranga river estuary is influenced by many factors including the quantitative variation of biogenic and organic substances (**Table 6**).

Correlation is a univariate statistical tool that is used to compute the rating scale of interrelation between two variables [37]. This interrelation rating scale was calculated through correlation analysis from the values of the regiment of water quality variables of the study area. The correlations between various hydrochemical parameters in the Auranga estuary are given in **Table 4**.

At the monitoring sections situated downstream of the wastewater discharge, high values of iron, phosphate, and nitrogen compounds have been identified, more exactly of the nitrate and ammonium ions, which influence the quality of the

| Parameters | Index/ rank | Relative index (Wi) | Downstream (Site 1) | | Upstream (Site 2) | |
|---|---|---|---|---|---|---|
| | | | Quality rate (Qi) | Sub index (SI) | Quality rate (Qi) | Sub index (SI) |
| Chloride | 1 | 0.0240 | 806.44 | 19.35 | 490.54 | 11.77 |
| Sulfate | 2 | 0.0487 | 280.11 | 13.64 | 223.00 | 10.86 |
| Alkalinity | 2 | 0.0487 | 192.67 | 09.38 | 107.75 | 05.24 |
| Iron | 2 | 0.0487 | 172.03 | 08.37 | 270.63 | 01.18 |
| Nitrate | 2 | 0.0487 | 151.80 | 07.39 | 127.00 | 06.18 |
| Color | 3 | 0.073 | 121.66 | 08.88 | 72.22 | 05.27 |
| DO | 3 | 0.073 | 111.20 | 08.11 | 74.40 | 05.43 |
| pH | 4 | 0.097 | 95.20 | 09.23 | 97.20 | 09.42 |
| Boron | 4 | 0.097 | 78.91 | 07.65 | 77.75 | 07.54 |
| COD | 4 | 0.097 | 57.35 | 05.56 | 59.30 | 05.75 |
| Temperature | 4 | 0.097 | 79.43 | 07.70 | 83.33 | 08.08 |
| Phosphate | 5 | 0.122 | 44.00 | 05.36 | 34.00 | 04.14 |
| TDS | 5 | 0.122 | 43.70 | 05.33 | 36.41 | 04.44 |
| WQI | | | | **115.97** | | **85.30** |
| PI | | | | **1.09** | | **0.86** |

**Table 6.**
*Annual water quality index and pollution index from downstream and upstream of Auranga estuary.*

watercourses. Water pollution by nitrate and phosphate reaches quite higher levels due to the introduction of compact procedures of farming, with it more utilization of chemical manures and more numbers of creatures in limited areas, especially in animal farming complexes from the Auranga hydrographical basin. During the analyzed period (2019) the evaluation of the quality status of estuarine watercourses, existing within the Auranga hydrographical system has revealed the fact that the river has been found in good water quality status. The tendency of the water quality index is assessed by the pecuniary activities in the agriculture, industrial, and residential areas in the sampling stations' vicinity in the Auranga hydrographical basin. For these reasons, constant monitoring is necessary, especially because this river flows further through the territory of Arabian Seawater.

## 4. Conclusion

The Water Quality Index of Auranga River Estuary ranges from 85.30 (Upstream) to 115.97 (Downstream) that indicates good water and moderate water quality respectively. The Pollution Index of water of Auranga Estuary ranges from 1.09 to 0.86 which indicates that the estuarine water is slightly polluted. The rapid industrialization and anthropogenic activities along the estuarine system and the coastal areas have brought a considerable decline in the water quality of the estuary. As per the bimonthly evaluation, contamination of components was higher in the months of November–December. Seasonal variation showed that most of the variables were more in contamination during the winter season. The water quality of Auranga estuary was slightly poor in the winter season due to more component loadings and good water quality during monsoon season because of more freshwater influence of heavy rain. Hence, the distribution pattern of nutrients in this estuary is controlled by many factors such as sewage from the industries, urban area, and agricultural sources, estuarine dynamics, fish processing unit, etc. During the analyzed period (2019) the evaluation of the water quality status of estuarine watercourses (rivers), existing within the Auranga hydrographical system has revealed the fact that the river has been found in good quality status. Awareness about anthropogenic pressure generated on water sources is necessary for the identification of the quality of water bodies and ultimately for adopting distinct methodology to protect and conserve the water in this region of Gujarat.

## Acknowledgements

The authors gratefully acknowledge SHODH, Govt. of Gujarat for financial support.

## Conflict of interest

The authors declare no conflict of interest.

## Author details

Shefali S. Patel* and Susmita Sahoo
N.V. Patel College of Pure and Applied Sciences, Gujarat, India

*Address all correspondence to: shefalipatel1312@gmail.com

## References

[1] Pritchard DW. In: Lauff GH, editor. What Is An Estuary: Physical Viewpoint. American Association for the Advancement of Science, US. 1967. pp. 3–5

[2] Gray JS. Marine biodiversity: Patterns, threats, and conservation needs. Biodiversity and Conservation. 1997;**6**(1):153-175

[3] Van Damme S, Struyf E, Maris T, Ysebaert T, Dehairs F, Tackx M, et al. Spatial and temporal patterns of water quality along the estuarine salinity gradient of the Scheldt estuary (Belgium and The Netherlands): Results of an integrated monitoring approach. Hydrobiologia. 2005;**540**(1–3):29-45

[4] Sun X, Zhang H, Zhong M, Wang Z, Liang X, Huang T, et al. Analyses on the temporal and spatial characteristics of water quality in a seagoing river using multivariate statistical techniques: A case study in the Duliujian River, China. International Journal of Environmental Research and Public Health. 2019;**16**(6):10-20

[5] Dunca AM. Water pollution and water quality assessment of major transboundary rivers from Banat (Romania). Journal of Chemistry. 2018; **2018**:1-18

[6] Niemi GJ, De Vore P, Detenbeck N, Taylor D, Lima A, Pastor J, et al. Overview of case studies on recovery of aquatic systems from disturbance. Environmental Management. 1990; **14**(5):571-587

[7] Russell IA. Spatio-temporal variability of surface water quality parameters in a South African estuarine lake system. African Journal of Aquatic Science. 2013;**38**(1):53-66

[8] Varol M, Gökot B, Bekleyen A, Şen B. Water quality assessment and apportionment of pollution sources of Tigris River (Turkey) using multivariate statistical techniques—a case study. River Research and Applications. 2012; **28**(9):1428-1438

[9] Manahan SE. The geosphere and geochemistry. In: Environmental Chemistry. 6th ed. Boca Raton, FL: Lewis; 1994. pp. 433–456

[10] Xu HS, Xu ZX, Wu W, Tang FF. Assessment and spatiotemporal variation analysis of water quality in the Zhangweinan River Basin, China. Procedia Environmental Sciences. 2012; **13**:1641-1652

[11] Kumar N, George B, Kumar RN, Sajish PR, Viyol S. Assessment of spatial and temporal fluctuations in water quality of a tropical permanent estuarine system- Tapi, West Coast India. Applied Ecology and Environmental Research. 2009;**7**(3):267-276

[12] Liu J, Zhang X, Xia J, Wu S, She D, Zou L. Characterizing and explaining Spatio-temporal variation of water quality in a highly disturbed river by multi-statistical techniques. Springer Plus. 2016;**5**(1):1171

[13] Jinwal A, Dixit S. Pre and post-monsoon Variation in physio-chemical characteristic in groundwater quality in Bhopal. India. Asian J. Exp. Science. 2008;**22**(3):311-316

[14] Chu HJ, Kong SJ, Chang CH. Spatio-temporal water quality mapping from satellite images using geographically and temporally weighted regression. International Journal of Applied Earth Observation and Geoinformation. 2018; **65**:1-11

[15] Islam MS, Tanaka M. Impacts of pollution on coastal and marine ecosystems including coastal and marine

fisheries and approach for management: a review and synthesis. Marine Pollution Bulletin. 2004;**48**(7–8):624-649

[16] Ionuş O. Water Quality Index-Assessment Method of the Motru River Water Quality (Oltenia, Romania). Annals of University of Craiova. Series Geography/Analele Universitatii din Craiova. Romania: Seria Geografie; 2010. p. 13

[17] Sivaranjani S, Rakshit A, Singh S. Water quality assessment with water quality indices. International Journal of Bioresource Science. 2015;**2**(2):85-94

[18] Kumar N, Kumar P, Basil G, Kumar RN, Kharrazi A, Avtar R. Characterization and evaluation of hydrological processes responsible for the spatiotemporal variation of surface water quality at Narmada estuarine region in Gujarat, India. Applied Water Science. 2015;**5**(3):261-270

[19] Vishnupriya SI. Assessment of coastal water quality through weighted arithmetic water quality index around Visakhapatnam, Bay of Bengal, India. Assessment. 2015;**4**(12):11775-11781

[20] APHA. Standard Methods for the Examination of Water and Wastewater Analysis. Washington DC: American Public Health Association, APHA-AWWA-WEF; 2005. ISBN: 0875532357 9780875532356

[21] Maiti SK. Handbook of Methods in Environmental Studies. Jaipur: ABD Publishers; 2004. p. 1

[22] Tyagi S, Sharma B, Singh P, Dobhal R. Water quality assessment in terms of water quality index. American Journal of Water Resources. 2013;**1**(3): 34-38

[23] Batabyal AK, Chakraborty S. Hydrogeochemistry and water quality index in the assessment of groundwater quality for drinking uses. Water Environment Research. 2015;**87**(7): 607-617

[24] Yan CA, Zhang W, Zhang Z, Liu Y, Deng C, Nie N. Assessment of water quality and identification of polluted risky regions based on field observations & GIS in the Honghe River watershed, China. PLoS One. 2015;**10**(3):1-13

[25] Gao S, Zhu R. Summary study on the development of water assessment. In: IOP Conference Series: Earth and Environmental Science. UK: IOP Publishing. 2018;**178**(1):012004

[26] Laura Ş, Silvica O, Anişoara I, Radu B. Quality Indices of the Water in the Middle Timiş River Basin. Annals of the University of Oradea. Romania: Fascicola Environmental Protection; 2013. p. 17

[27] Fatema K, Maznah WW, Isa MM. Spatial and temporal variation of Physico-chemical parameters in the Merbok Estuary, Kedah, Malaysia. Tropical Life Sciences Research. 2014; **25**(2):1-19

[28] Bortoletto EC, Silva HA, Bonifácio CM, Tavares CR. Water quality monitoring of the Pirapó River watershed, Paraná, Brazil. Brazilian Journal of Biology. 2015;**75**:148-157

[29] Yan CA, Zhang W, Zhang Z, Liu Y, Deng C, Nie N. Assessment of water quality and identification of polluted risky regions based on field observations & GIS in the honghe river watershed, China. PLoS One. 2015;**10**(3):e0119130

[30] Zhang L. Different methods for the evaluation of surface water quality: The case of the Liao River, Liaoning Province, China. International Review for Spatial Planning and Sustainable Development. 2017;**5**(4):4-18

[31] Shrestha S, Kazama F. Assessment of surface water quality using multivariate statistical techniques:

A case study of the Fuji river basin, Japan. Environmental Modelling & Software. 2007;**22**(4):464-475

[32] Varol M, Şen B. Assessment of surface water quality using multivariate statistical techniques: A case study of Behrimaz Stream, Turkey. Environmental Monitoring and Assessment. 2009;**159**(1):543-553

[33] Fu X, LihuaTeng XW, Lv J, Zhou Y, Tang X, Qian J, et al. Research article application of fuzzy comprehensive method in water quality assessment of the Yongjiang River, China. International Journal of Fisheries and Aquatic Sciences. 2016;**5**(1):1-6. DOI: 10.19026/ijfas.4.2984

[34] Li R, Zou Z, An Y. Water quality assessment in Qu River based on fuzzy water pollution index method. Journal of Environmental Sciences. 2016;**50**: 87-92

[35] Milijašević-Joksimović D, Gavrilović B, Lović-Obradović S. Application of the water quality index in the Timok River basin (Serbia). Journal of the Geographical Institute" Jovan Cvijic", SASA. 2018;**68**(3):333-344

[36] Alves JD, Fonseca LC, Chielle RD, Macedo LC. Monitoring water quality of the Sergipe River basin: An evaluation using multivariate data analysis. RBRH. 2018;**10**:23

[37] Singh JV. Performance, slack, and risk taking in organizational decision making. Academy of Management Journal. 1986;**29**(3):562-585

Chapter 5

# On the Design of Total Water Use-Based Incentive Schemes for Groundwater Management

*Wided Mattoussi, Mohamed Salah Matoussi and Foued Mattoussi*

## Abstract

Groundwater over-pumping by manipulating water meters may constraint the efficient use of the resource, leading to the potential aquifers' deterioration. Well designed institutional arrangements might be effective at reducing over-exploitation. The objective of this research was to shed light on the design of various incentive schemes to face groundwater over-pumping ranging from individual water use-based incentive schemes, where individual withdrawals are the users' private information, to total water use-based incentive schemes, where the aggregate withdrawal is publicly observable. For the latter setting, two schemes were proposed. The first one is within the framework of moral hazard in teams, where the Water Authority administers monetary incentives that do not balance the budget, restoring thereby the full-information outcome. The second scheme promotes a cooperative management governed by a collective responsibility rule that induces peer monitoring by members. We show that groundwater overuse is more likely when monitoring costs are high, punishments are weak and cooperatives are large. We also show how the cooperative size and punishments are determined endogenously by constraints on monitoring. We extend the basic analysis to study collusion in monitoring between cooperative members and compare different monitoring structures. The results confirm that well-designed incentives and institutions can reduce groundwater over-exploitation, and that constraints on monitoring costs affect institutional design.

**Keywords:** groundwater over-pumping, moral hazard in teams, cooperatives, peer monitoring, cooperative size, collusion, monitoring structure

## 1. Introduction

Groundwater resources are important for at least two reasons: Firstly, they are well appropriate for drinking[1] due to their (generally) high quality [2, 3]. Secondly, groundwater reservoirs constitute very important long-term storage [4–6], particularly useful during long periods of droughts characterizing arid and semi-arid regions.

[1] Falkenmark [1] estimates that one-third of the world's population rely on groundwater supply for drinking.

The two major threats to groundwater are over-exploitation and pollution [7]. The rapid demographic growth, urbanization, industrialization, intensification of farming practices and climate change pressures have led to an increasing demand for groundwater as a reliable source of water supply.[2]

Intensive abstraction can deplete the groundwater in an aquifer. It might be possible to over-pump any aquifer temporary during periods of droughts, but durable over-exploitation would certainly lead to irreversible degradations [9], due for example to saline intrusion, sea water intrusion and quality deteriorations generated by declining water tables. The artificial recharge of aquifers is appealing, but it might not be implemented on a large scale, meaning that only the long-term natural replenishment guarantees the groundwater conservation. The only course left open is then a better management of the resource.

A large set of policy instruments have been developed for a better groundwater management, including economic instruments and institutional arrangements. Several economic instruments are used to limit over-exploitation such as water quotas, pumping taxes or marketable pumping permits. They however focus on individual withdrawals which are assumed to be publicly observable. This is rarely the case in the real world as legal and administrative settings are generally insufficient to perfectly monitor individual withdrawals, meaning that groundwater may present open-access resource features [10]. Groundwater is hence withdrawn in an imperfect informational context and the above instruments become ineffective.

Economists now agree on the fact that resource allocation in less developed economies is profoundly influenced by non firm institutions such as group lending, credit cooperatives, sharecropping, Water Users Associations and so forth. In developing countries various water institutions coexist. They range from centralized regulation, where management responsibility is entirely delegated to government agencies, to markets[3] for tradable water rights where farmers can sell their water shares to higher value uses. In between lies the entire spectrum of water allocation methods characterized by levels of decentralization, including Water Users Associations (referred to as water cooperatives), where users are involved in the decision-making process, and are thereby entrusted with part of the management responsibility normally held by government agencies. Water institutions can be effective at improving the resource allocation whenever they are well designed. Institutions influence individual behavior through incentives they give rise to, but institutions themselves evolve endogenously in part because of their incentive properties [11].

This paper sheds light on the design of various incentive schemes to face groundwater over-exploitation by farmers who can over-pump water typically by manipulating their individual water meters in an asymmetric information context.[4] In the sequel of this work, we refer to groundwater over-pumping as groundwater theft. The study in particular shows how the effective design of water institutions in response to a perceived problem of theft can help to reduce theft of water, and thereby improve water use efficiency. The response of the Water Authority (hereafter, WA), to tackle theft will differ according to whether it uses an incentive

[2] Groundwater constitutes about 89% of the freshwater on our planet (discounting that in the polar ice caps) [8] (p.1).

[3] The most celebrated case of tradable water rights comes from Chile, where agrarian reforms and the Water Code of 1981 formalized water rights, and allowed for water sales separately from sales of land.

[4] Despite the relevance of asymmetric information problems for water management, only a few studies examine the application of such concepts to irrigation water in general and to irrigation groundwater in particular [12].

scheme based on the individual farmer's withdrawal which is her private information, namely the centralized management which dominated water management in developing countries for several decades [13] or it resorts to total water use-based incentive schemes, where the total amount of water withdrawn by farmers is publicly observable and payments from farmers can be conditioned on it.

In the centralized scheme, farmers who steal, do so directly from the WA, and thereby do not impose negative externalities on each other. The WA tries to reduce theft by directly monitoring the farmers' behavior, punishing observed instances of theft. We show in the model that some monitoring is always required in equilibrium. The WA tolerates some theft in order to save in monitoring costs.

In the decentralized management, where unobserved farmers withdrawals can be regulated through instruments based on collective performance (observed aggregate withdrawals) two schemes will be proposed. The first one corresponds to the framework of moral hazard in teams' problem where the WA administers incentive schemes that do not balance the budget, restoring thereby the full-information outcome [14]. Such scheme works independently of the team size, but it may be infeasible when farmers have endowment constraints. This is why one may resort to an alternative team-based incentive scheme that might not violate individual endowment constraints and in which the WA makes use of the informational advantages farmers have over the WA because of their long standing and high trade links (especially in close-knit societies). This corresponds to cooperative institutions which are very likely to be well suited to deal with a variety of collective action problems associated with water management, though their success in doing so depends on some particular features of their design. In this research we show that such institutions may also be well suited to dealing with groundwater theft; we discuss the features of their design that enable them to do so. We show that the incentives for theft vary considerably in response to these features and discuss implications for policy. We in particular consider the properties of cooperatives characterized by a collective responsibility rule, which makes all members jointly liable for aggregate withdrawals, and show that this feature is likely to induce peer monitoring by cooperative members[5] which is likely to be more effective at reducing theft than any means available to more centralized structures. We in particular show that groundwater theft is more likely when monitoring costs are high, punishment levels are weak and cooperatives are large. Moreover, straightforward comparison of the two team-based schemes shows that with sufficiently stringent punishments, the two schemes achieve the full-information water use level. Otherwise, theft occurs in cooperatives and a positive monitoring effort is required in equilibrium, meaning that the first team-based incentive scheme outperforms the cooperative management.

The results in the cooperative setting are obtained for given levels of punishment and cooperative size, but cooperatives are typically able to influence both of these variables. The model shows that these institutional characteristics are endogenously determined by constraints on monitoring: Higher monitoring costs increase punishment levels and reduce the cooperative size. Simulations also show that cooperatives can be neither too small because of the "monitoring costs savings" effect nor too large because of the "stealing" effect.

We extend the analysis thereafter to tackle the issue of collusion in monitoring efforts of cooperative members and show that collusion is welfare enhancing. We then compare among different monitoring structures, mutual and localized monitoring. Although in practice the mutual monitoring structure - whereby each farmer

[5] There is now a substantial literature on peer monitoring [15–21].

in the cooperative is being simultaneously monitored by all of her peers - is commonly observed [15], other monitoring structures deserve consideration. An interesting departure from the mutual structure is the "*localized monitoring*" in which every farmer monitors (and is monitored) by only one of her peers avoiding therefore the duplication of monitoring. We show that in equilibrium the localized monitoring effort is higher than twice the mutual monitoring level. This result is driven by the distributional character of peer monitoring, where monitoring allows cooperative members to shift the cooperative fine on others in addition to reducing theft.

The paper is organized as follows. In Section 2, we review the relevant literature. Section 3 sketches our model. In Section 4, we present an individual water use-based incentive scheme. In Section 5, we propose two total water use-based incentive schemes. We state a number of propositions describing the dependence of groundwater theft on a number of determinants, some of which are themselves determined by more fundamental factors including costs of monitoring. Section 6 provides two extensions of the basic model: the analysis first allows for collusion in monitoring efforts of cooperative members and then compares different monitoring structures. Section 7 proposes some policy recommendations. In Section 8, we present some potential extensions for further research. Section 9 concludes. Mathematical details are relegated to an Appendix.

## 2. Literature review

Our research relates to various types of literature emphasizing the team nature of a problem. It relates to the peer monitoring on group lending programs and to decentralized groundwater management literatures, where peer monitoring is recognized as an effective instrument in mitigating the moral hazard behavior of individuals linked by a collective responsibility rule. It also relates to the non-point source pollution and to the non-point groundwater withdrawals, where unobserved individual emissions (withdrawals) can be regulated through instruments based on collective performance, which is the level of observed aggregate (ambient) pollution (withdrawals).

In the peer-monitoring literature, peer monitoring is an important means to mitigate free riding in groups of borrowers related by a joint-liability clause that creates an incentive mechanism in which each group member has an interest in screening and monitoring the other members. In the case of non-repayment by the group, all members will be denied future access to loans from the program, and defaulters who are caught may face fixed social sanctions. The seminal publications in this area are Stiglitz [19] and Varian [21], who show that the (costless) peer monitoring within groups can prevent members' shirking in their productive efforts (Varian), and reduce poor project selection (Stiglitz), improving thereby repayment rates and reducing the costs of lending. More recently several papers including Ghatak and Guinnane [20], Armendariz [15], Che [17], and Conning [18], elaborate on the Stiglitz-Varian models, relax the assumption of the costless peer monitoring and deal with various extensions including the optimal group size, monitoring structures and the dynamic aspect of contractual relationship between group members.

In the context of decentralized groundwater management, Montginoul et al. [22] mentioned the use of a mechanism that consists of providing incentives for all groundwater users getting involved in the monitoring of groundwater abstraction to monitor each other, in order to increase the probability of control. The cost of decentralized monitoring (peer monitoring) is expected to be lower, since agents have more information on the actions of other agents (areas and crops irrigated, irrigation practices and frequencies, etc.) than any centralized structure. The incentive to participate in such a decentralized monitoring system can be provided by redistributing a share of the fine to the person who discovers the defaulter. This system has been used for centuries for regulating access to forests and common pastures in the Italian Alps [23]. This mechanism is likely to be rejected in many cultural contexts as it may be strongly assimilated to denouncement.

The literature on non-point source pollution follows the pioneering work of Segerson [24], whose analysis built on the earlier theoretical analysis of Holmström[6] [14], who addressed the problem of free riding in teams in a more general environment.

Segerson [24] proposed a target based mechanism (TBM) where a regulator should monitor ambient pollution concentrations of a well-defined group imposes to each group member a tax (or a subsidy) proportional to the difference between observed group emission level and the group target. She shows that for a sufficiently high level of the ambient tax, the Nash equilibrium yields an aggregate pollution level equal to the group target.

Segerson's work has inspired several intriguing extensions (e.g., Xepapadeas [25]; Miceli and Segerson [26]; Karp [27]; Millock and Salanié [28]). Xepapadeas [25] proposed a scheme of subsidies and two fining regimes: collective and random fining. Under collective fining, all firms are charged a fine whenever the observed ambient pollution level lies above some predetermined standard. Under the random fining scheme, only one firm is randomly chosen to be punished, irrespective of being responsible for the whole group's deviation from the standard level.

Miceli and Segerson [26] proposed the introduction of collective responsibility rules among group members that create incentives similar to the ones created by ambient taxes. However, Litchenberg [29] noted that these liability rules are not likely to be first-best and are probably best-suited for controlling pollution related to the use of hazardous materials, or for non-frequent occurrences of environmental degradation like oil spills. Karp [27] suggested a model in which polluting firms behave strategically with respect to the regulator and found that their tax burden is lower under an ambient tax than taxes based on individual emissions, provided that the tax adjusts quickly, firms are patient, and the number of firms is small. Millock and Salanié [28] proposed a model of ambient taxes, where polluters might cooperate, and show that ambient taxes give strong incentives towards cooperation. However, when the degree of cooperation is unknown, the optimal regulation requires the regulator to offer a choice between a standard Pigouvian tax and a much lower ambient tax.

Although theoretically appealing, ambient-based schemes are rarely implemented in the field (an exception is presented in Ribaudo, Horan, and Smith [30]) for numerous technical, practical and political reasons [31].

The "ambient tax" instrument proposed by economic literature to solve diffuse pollution problems can be well suited to manage unobserved groundwater withdrawals since withdrawals of a well-defined group can be approximated by the groundwater table level monitored at some observation points [9]. The ambient tax (subsidy) would be charged (paid) to all users if the groundwater table falls below (above) the target level set by a regulator which was decided to not overpass.

In the decentralized management of groundwater several authors including Giordana [32], Lenouvel et al. [12] and Figureau et al. [33], show that contract-based instruments may play a significant role in reducing groundwater over-pumping.

[6] One main finding of Holmström is that in the absence of uncertainty, no budget balancing mechanism exists to solve the problem for avoiding individual free riding in teams.

Giordana [32] proposed three incentive instruments to fight groundwater over-pumping: a tax/subsidy over reported individual withdrawals with a random audit and penalties in case of non-compliance by groundwater users, an ambient tax, and a mixed instrument combing both instruments. He shows that the latter scheme outperforms the former two schemes.

Lenouvel et al. [12] proposed an optional target-based mechanism to reduce groundwater over-exploitation when farmers' behavior is imperfectly monitored. The mechanism combines a classical ambient tax, paid by all farmers of the area when the water table level falls below a pre-defined target, with an optional individual contract signed with the regulator in which signatory farmers commit to provide true information to the regulator concerning the location of their wells, irrigated fields, and volume pumped, and to facilitate the control of this information. These farmers avoid the collective sanction if they comply with an individual quota. This mechanism is tested experimentally in the lab with a contextualized protocol and results show that it reduces withdrawals but that subjects are able to coordinate in a repeated setting to extract an informational rent.

Figureau et al. [33] have proposed three policy instruments, which can be used to enhance farmers' compliance with individual groundwater allocations for irrigation in a decentralized management context. The first policy couples economic incentives by combing the use of a penalty with a reward. The penalty consists of a tax charged to farmers who exceed their allocation and is proportional to the over-pumping. The revenues from this penalty system are then shared between farmers who withdraw less than their entitlement, each one receiving a share proportional to their water saving.

The second policy is a "pooling agreement" through which farmers would agree to mutualize their quotas, in the sense that some farmers agree to relinquish part of their individual water allocation to help other farmers confronted by unusual situations. The volume given back to the Groundwater Users Association (GWUA) is then redistributed to farmers who have an exceptional need for extra water. The internal redistribution follows general principles and rules, which have been validated by the farm community. The contract is favorable to the agents as a team relative to the standard penalty system provided that the team does not exceed the targeted abstraction level, but unfavorable to the team if the target is exceeded. The third policy combines payments and fines. Farmers exceeding their quota pay an increasing block fine for the extra water pumped. Revenues from fines are then redistributed among farmers who use less than their quota; the amount received by farmers is proportional to their water saving efforts.

The three policies are tested through experiments with farmers and results reveal a preference for the third scheme that combines economic and social incentives, as it is expected to meet water and budget balance simultaneously.

Our cooperative model differs from most of the existing theoretical literature on peer monitoring in two respects. First, in their models the punishment is fixed: in the case of non-repayment by the group, all members will be denied future access to loans from the program, and defaulters who are caught may face fixed social sanctions. However, in our model the punishment depends continuously on the extra water pumped by farmers. Second, peer monitoring in this paper is quite specific in that farmers are competing in monitoring, which gives rise to a distributional effect in addition to an incentive effect. Indeed, peer monitoring may allow each cooperative member to shift the cooperative fine on others in addition to mitigating the moral hazard behavior of group members.

As for the ambient tax literature, it differs from the cooperative model in several respects. First, in our model the joint liability clause creates incentives for peer monitoring by group members, while ambient taxes do not. Second, in the ambient tax mechanism each individual is taxed according to the socially marginal damage when ambient emissions deviate from some predetermined level of emission. In our study, however, the distribution of the punishment burden is endogenously determined by peer monitoring. Third, in our study whether efficiency is obtained or not depends on the stringency of the punishment burden. However, most mechanisms suggested in the ambient tax literature are theoretically suitable for implementing the efficient allocation of abatement efforts in a Nash equilibrium.

In the decentralized groundwater management using mechanisms creating incentives for peer monitoring, such incentives are created through joint liability clauses based on rewards rather than punishment sharing rules as in our model.

Our cooperative model differs from decentralized groundwater management using contract-based instruments in three respects. Firstly, their mechanisms do not create incentives for peer monitoring as in our model. Secondly, in our model there are no rewards for farmers using less than the allocated quota. Third, in their models, they mainly use economic instruments as rewards and punishments to mitigate aquifer over-exploitation, but they do not come close to institutional design of the group of groundwater users such as the optimal size of the group and the optimal punishment or reward.

## 3. The problem

Consider two identical farmers who pump water from a *shared renewable* aquifer to produce a homogeneous farm good. Even in this restricted setting certain features emerge that we believe to be interesting for policy implications and might be relevant for empirical investigations. Suppose that the yield $(y)$ response to water $(q)$ can be described by the relation $y = g(q)$, where $g(.)$ is increasing and concave. Each farmer bears a cost $c$, measured in units of output for every unit of water used. In addition, the farmer pays a linear price $p$ per unit of water, which is set by the Ministry of Agriculture. The quantity of water maximizing the farmer's profit equates the marginal value product of water to the marginal cost of generating such a quantity

$$q^{fi} : g'(q) = c + p. \tag{1}$$

The superscript $^{fi}$ refers to the full-information setting. In the complete information setting and when we abstract from any shadow cost of public funds that might imply Ramsey-pricing considerations, the WA can implement the first-best efficient outcome by setting $p$ equal to $\delta$. $\delta$ represents the full public cost of mobilizing water to irrigated areas, which includes investment costs, operation and maintenance costs, extraction externalities associated with pumping from a shared aquifer, and any shadow cost associated with the scarcity of water.

The farmer is allocated a *quota* equal to her full-information water use[7], $q^{fi}$. When the farmer's water use is her private information (unlike the aggregate amount of water used by all farmers which is publicly observable), the farmer who has an individual water meter can well exceed her quota by manipulating her meter. The amount of water used in excess (referred to as goundwater theft) can be written as $\alpha = q - q^{fi}$.

[7] It is hard to understand why the WA as a social planner should ever choose any quota but the full-information water use level.

The response of the WA will differ according to whether there is centralized management or management based on total water use which is publicly and costlessly known. We consider these two cases in turn.

## 4. Centralized scheme

Here we present the centralized management in an incomplete informational context. In what follows, a few assumptions necessary to the analysis are listed.

- *Assumption 1*: The WA invests in monitoring devices aiming at making individual withdrawals observable. Monitoring incurs a social cost denoted by $\Psi(m)$, which is increasing, convex and satisfies $\Psi(0) = 0$. The cost of monitoring should be understood as including not only measurement devices, but other costs such as the wages of monitors.

- *Assumption 2*: If the farmer is not monitored, then she pays the mandated water fee associated with her allotment, $pq^{fi}$. Otherwise, she is discovered exceeding her quota (stealing) with a probability $\pi(m)$ which increases with the intensity of monitoring. To simplify the exposition the probability $\pi(.)$ is assumed to be commonly known and takes the form

$$\pi(m) = \min\{\kappa m, 1\}. \tag{2}$$

where $\kappa > 0$ (we assume henceforth that it is sufficiently small to generate an interior solution, which is realistic).

- *Assumption 3*: When the farmer is detected stealing, her individual withdrawal is established without error and she pays $pq^{fi}$ plus a penalty, $F^{cs}$ proportional to the level of theft. The punishment is a monetary transfer from the farmer to the WA and takes the form

$$F^{cs} = f\max\{\alpha, 0\}. \tag{3}$$

where the punishment rate $f$ is positive, greater[8] than $p$ and given outside the model. There are no rewards for using less than the allocated quota. The solutions to the centralized and cooperative managements will be indexed with the superscript $^{cs}$ and superscript $^{c}$, respectively.

- *Assumption 4*: Let $\tilde{Q}$ be the aquifer storage capacity or the stock of the resource in situ. For feasibility requirements we assume throughout that the price of water is lower than the marginal yield of using half of the storage capacity from which is deducted the private cost of one unit of water

$$p \leq g'\left(\frac{\tilde{Q}}{2}\right) - c. \tag{4}$$

---

[8] Because otherwise the farmer will always have an interest in stealing everything. The net return from theft of water is equal to $(p - \kappa mf)\alpha$, which occurs with the probability $\kappa m < 1$. If $f < p$, one gets $\kappa mf < f < p$, and therefore theft is strictly beneficial; this essentially implies that the net return is maximized when the farmer steals everything.

Rewriting this condition yields $2(g')^{-1}(c+p) \leq \tilde{Q}$ (where $(g')^{-1}(c+p) = q^{fi}$), which states that the total quota allocated to farmers must be lower than the storage capacity. This is to guarantee a rate of utilization that does not exceed the rate of replenishment to avoid the depletion of aquifers.

The order of events is that the WA chooses monitoring[9], $m$ then each farmer chooses the quantity of water to use $q^{cs}$. In what follows we focus on the subgame perfect equilibrium and solve the model by backward induction. In stage 2 of the game, the farmer chooses $q^{cs}$ so as to maximize her expected payoff:

$$\max_q U^{cs}(q; m) = g(q) - cq - pq^{fi} - \kappa m f\left(q - q^{fi}\right).$$

Whose first-order condition is

$$g'(q^{cs}) = c + \kappa m f. \tag{5}$$

The performance of the centralized management relative to the full-information setting depends on the intensity of monitoring as summarized by corollary 1:

COROLLARY 1:

1. *If* $m < \frac{p}{\kappa f}$, *then* $q^{cs} > q^{fi}$;

2. *If* $m \geq \frac{p}{\kappa f}$, *then* $q^{cs} = q^{fi}$.

*Where* $\frac{p}{\kappa f}$ *is the minimum required level of monitoring that deters theft completely.*

The full-information outcome obtains if the farmer is intensively monitored; Otherwise, theft occurs, as it becomes a privately profitable activity, i.e., the expected net benefit from stealing is $(p - \kappa m f)\left(q^{cs} - q^{fi}\right) > 0$.

Now let us turn to the initial contracting stage, where the WA anticipates the farmer's behavior and picks monitoring, $m$ that maximizes the social welfare function. Specifically this function is the sum of the farmers' surpluses $2\left[g(q^{cs}) - cq^{cs} - pq^{fi} - \kappa m f\left(q^{cs} - q^{fi}\right)\right]$ and the water supplier surplus equal to the revenue from the expected payments for water use, $2\left[pq^{fi} + \kappa m f\left(q^{cs} - q^{fi}\right)\right]$ from which is deducted the cost of mobilizing the resource to the irrigated area, $2\delta q^{cs}$ and the cost incurred by monitoring, $2\Psi(m)$

$$W^{cs}(m, f) = 2\left[g(q^{cs}) - (c+\delta)q^{cs} - \Psi(m)\right]. \tag{6}$$

The WA must also consider two major constraints. The first one is the water availability constraint

$$2q^{cs} \leq \tilde{Q}. \tag{7}$$

which reflects the scarcity of the resource: farmers can at most use what is available. And the second one is the replenishment constraint

$$2q^{fi} \leq \tilde{Q}. \tag{8}$$

---

[9] The WA is able to control the punishment rate, $f$ in addition to controlling the monitoring decision. When punishment is endogenous and costly, some level of punishment is always required in equilibrium. However, because punishment is costly, the optimal response of the WA is to tolerate some theft of water in order to save in punishment costs.

which states that the rate of groundwater utilization should be lower than the replenishment rate. In what follows proposition 1 characterizes the solution to the WA's problem:

**PROPOSITION 1:** *Suppose that assumptions (1), (2), (3) and (4) hold and constraints (C1) and (C2) bind*[10]*, then the optimal monitoring $m^{cs}$ solves*

$$m^{cs} : h\frac{\kappa f}{g''(q^{cs})} = \Psi'(m^{cs}). \tag{9}$$

*where $\mu$ is the Lagrangian multiplier on constraint (C1) and $h = (\kappa m^{cs} f - \delta - \mu)$.*

**Proof**: See the appendix.

The proposition reveals that some monitoring is always required in equilibrium. However, because monitoring is costly, the optimal response of the WA would be to tolerate some theft in order to save in monitoring costs. Moreover, the equilibrium monitoring effort responds directly to the degree of water scarcity, captured by parameter $\mu$ (the scarcity rent or shadow value of water).

$$\frac{\partial m^{cs}}{\partial \mu} = \frac{\kappa f}{g''(q^{cs})} \frac{1}{\left[\frac{(\kappa f)^2}{g''(q^{cs})} - h\frac{g'''(q^{cs})}{g''(q^{cs})} - \Psi''(m^{cs})\right]} > 0. \tag{10}$$

The more severe the shortage of water is, the higher is the required monitoring effective at reducing theft.

## 5. Team-based incentive schemes

### 5.1 First scheme

We assume that the aggregate water use of the two farmers, $Q = q_1 + q_2$, is publicly and costlessly known, and can be contracted for directly. In particular the WA designs a team-based incentive scheme where it asks the farmer to pay the fixed water fee associated with her allocated quota, $pq_i^{fi}$ and a share of the full extra amount if actual water use exceeds the total quota allocated to the group, $s_i\left[p.\left(Q - 2q_i^{fi}\right)\right]$ for $i = 1, 2$, where $s_i(.)$ is differentiable. We use a very restricted strategy set for the agents, and we shall look only at symmetric equilibria which implies that

$$q_i^{nbb} = q^{nbb} \text{for} i = 1, 2.$$

and that every team member is allocated the same quota $q_i^{fi} = q^{fi}$ for $i = 1, 2$. The solution to this problem is indexed with the superscript $^{nbb}$, referring to *non balanced budget*.

The order of events is that taking the price of water[11] $p$ for given, each farmer chooses the quantity of water to use $q^{nbb}$ that maximizes her expected payoff

$$\max_q U(q) = g(q) - cq - pq^{fi} - s_i\left[p.\left(Q - 2q^{fi}\right)\right].$$

---

[10] Since $q^{cs} \geq q^{fi}$, then (C2) is only binding if (C1) is also binding, and so we can ignore (C2).

[11] Which is set by the Ministry of Agriculture.

Whose first-order condition is

$$g'(q) = c + s_i'\left[p.\left(Q - 2q^{fi}\right)\right].p, \tag{11}$$

A mere comparison of (1) and (11) gives

$$s_i'\left[p.\left(Q - 2q^{fi}\right)\right].p = p,$$

and thereby

$$s_i'\left[p.\left(Q - 2q^{fi}\right)\right] = 1,$$

Implying that each farmer has to pay the total liability, $p.\left(Q - 2q^{fi}\right)$ to the WA.

$$s_i\left[p.\left(Q - 2q^{fi}\right)\right] = p.\left(Q - 2q^{fi}\right). \tag{12}$$

The WA can restore the full-information outcome by administering incentive schemes that *do not balance the budget*[12] since both farmers will be paying the full extra amount $p.\left(Q - 2q^{fi}\right)$.

This incentive scheme works independently of the team size, but its implementation may be constrained by the farmers' limited liabilities. That is why the WA may promote cooperative behavior.

### 5.2 Cooperative management

Similarly to the previous scheme we assume that the total water use by cooperative members, $Q = q_1 + q_2$, is publicly observable, and aggregate payments from the cooperative to the WA can be conditioned on it. In particular, this feature allows for a collective responsibility rule: when theft occurs, the cooperative as a whole receives a punishment proportional to the total amount of water stolen:

$$F^c = f\left(\sum_{i=1,2} q_i - q_i^{fi}\right). \tag{13}$$

Now suppose that, relative to the WA, farmers have informational advantages in monitoring each other, as a result of social ties and/or spatial proximity and neighborhood and/or long term trade relations.

We assume that peer monitoring brings about only evidence of the occurrence theft but not of its amount.[13] The WA may then contemplate the possibility of inducing peer monitoring between cooperative members by setting a collective responsibility rule that makes all members jointly liable: If theft occurs, the fine inflicted on the cooperative as a whole is shared equally between farmers who are caught stealing; otherwise it is shared by all members.

---

[12] If the WA has instead administered the incentive scheme where the total liability $p.\left(Q - 2q^{fi}\right)$ was fully shared among the agents, this would result in an inefficient outcome [14]. The point is therefore not that group punishments is the only effective scheme, but rather budget-breaking is the essential instrument in neutralizing the free-riding problem.

[13] All what a farmer may observe is whether the other cooperative members manipulate their meters and if they do, the evidence about these actions will be established with certainty. Indeed, only the farmer herself and the WA can have access to the farmer's water meter.

Performing peer monitoring is costly, we denote by $\psi(m)$ the associated cost, which is assumed to be increasing, convex and satisfies $\psi'(0) = 0$. Each farmer commits to a level of monitoring[14] (observable by the other members) before the choice of the level of water to use is made. The probability that a farmer $i$ is caught stealing water is then given by:

$$\pi_i(m_j) = \min\{\kappa m_j, 1\}, \tag{14}$$

where $\kappa > 0$. Farmers do not collude in either their production or monitoring decisions.[15]

The order of events is that taking for given the price of water $p$, cooperative members choose $m_i$, then having observed each others' choice of $m_i$ they choose the level of water to use $q_i$. Suppose that both farmers steal water, i.e., $\alpha_k > 0$ for $k = i, j$. The expected share[16] of farmer $i$ from the cooperative fine is decreased by her peer's monitoring, and is in turn increased by her own monitoring.

$$s_i^{\exp} = \frac{1}{2}(1 - \kappa m_i + \kappa m_j). \tag{15}$$

The subgame perfect equilibrium is the profile $(m_1^c, m_2^c, q_1^c, q_2^c)$ of monitoring efforts $m_i^c \geq 0$ and water use levels $q_i^c$ mapping from the set of monitoring decisions into the set of water use decisions, $q_i^c : [0, +\infty)^2 \to 0, \tilde{Q})$. In what follows we shall focus on *symmetric* equilibria which imply that

$$q_i^c = q^c \text{ and } m_i^c = m^c \text{ for } i = 1, 2.$$

and that every cooperative member is allocated the same quota $q_i^{fi} = q^{fi}$ for $i = 1, 2$.

Similarly to the centralized structure, in cooperatives we shall restrict attention to the punishments that are higher than the price of water, i.e.,$f > p$, because otherwise farmers will always have an interest in stealing everything. The outcome will depend on the stringency of the punishment rate. If it is sufficiently high, i.e., when $f$ lies above $2p$, farmers will neither steal, nor monitor in equilibrium (as high punishments ensure that the collective penalty will be severe enough to deter theft). Otherwise[17], i.e., when $f \in (p, 2p)$, farmers will steal and monitor in equilibrium. Summarizing:

---

14 One may think of observable sunk investments (such as tools and equipment) being made by members of the cooperative, and which would commit them to a higher monitoring effort. For instance, it is widely observed in developing countries like Tunisia that landlords build small houses in their farms where they can keep some farm equipment for daily use and where both landowners and agricultural laborers may spend some time.

15 For the moment we sidestep the issue of collusion in monitoring efforts, but we return to it later in Section 7.1.

16 The expected share of farmer $i$ from the cooperative fine when everyone steals water is given by

$$s_i^{\exp} = \frac{1}{2}\kappa m_i \kappa m_j + \kappa m_j(1 - \kappa m_i) + \frac{1}{2}(1 - \kappa m_i)(1 - \kappa m_j)$$

Where the first term corresponds to her share when both farmers are caught stealing, the second term is her share when she is caught and farmer $j$ not, and the last term is her share when none is caught. Rewriting the expression above yields the expression (15) in the text.

17 It is worth noting that focusing on this range of punishment levels is less restrictive than it seems. Indeed, such a restriction holds only for the two-farmer cooperative; for the more general case of n-farmer cooperative, punishment rates will instead strictly lie between $p$ and $np$.

**PROPOSITION 2:** *If $f \geq 2p$, there exists a unique symmetric subgame perfect equilibrium $(q^c, m^c)$ such that*

$$q^c = q^{fi} \text{ and } m^c = 0. \tag{16}$$

*If $p < f < 2p$. Then, the unique symmetric subgame perfect $(q^c, m^c)$ satisfies*

$$q^c : g'(q) = c + \frac{1}{2}f, \tag{17}$$

and

$$m^c = \phi(k_2). \tag{18}$$

*Where, $k_2 = f\left[(q^c - q^{fi}) - \frac{1}{4}\frac{f}{g''(q^c)}\right]$ and $\phi = (\psi')^{-1}$.*

**Proof:** See the appendix.

Peer monitoring reduces groundwater theft[18] (*incentive effect*) and it may allow every cooperative member to shift[19] the cooperative fine on the others (*distributional effect*).

The immediate corollary of proposition 2 is that the comparative evaluation of the performance of the two team-based incentive schemes depends on the stringency of punishment rates. For sufficiently stringent punishments (i.e., $f \geq 2p$), the incentive problem is eliminated altogether and the full-information outcome obtains. Otherwise, theft occurs in cooperatives and a positive monitoring effort is required in equilibrium; consequently, the first team-based incentive scheme outperforms the cooperative management. The following corollary states this point.

**COROLLARY 2:**

1. *If $f \geq 2p$, then*

$$q^c = q^{nbb} \text{ and } W^c = W^{nbb}, \tag{19}$$

2. *If $p < f < 2p$, then*

$$q^c > q^{nbb} \text{ and } W^c < W^{nbb}. \tag{20}$$

*Where, $q^{nbb} = q^{fi}$ and $W^c = 2[g(q^c) - (c+\delta)q^c - \psi(m^c)]$ and $W^{nbb} = W^{fi} = 2\left[g(q^{fi}) - (c+\delta)q^{fi}\right]$ are the cooperative and full-information social welfare levels, respectively.*

### 5.3 Comparative statics

To obtain explicit solutions where possible we assume that monitoring costs take the quadratic form $\psi(m) = \frac{1}{2}bm^2$ where $b > 0$. We first investigate the impact of

---

[18] The partial derivative of the level of theft undertaken by farmer $i$ with respect to monitoring performed by her peer, $m_j$ is given by

$$\frac{\partial(q_{i-}q^{fi})}{\partial m_j} = \frac{\kappa f}{2g''(q_i)} < 0.$$

[19] This finding follows from equation (15). The expected share of farmer $i$ from the total fine increases with the monitoring effort performed by her peer, $m_j$:

$$\frac{\partial s_i^{\exp}}{\partial m_j} = \frac{1}{2}\kappa > 0.$$

water price and punishment on the equilibrium level of theft. As one intuitively expects, theft increases with water price and decreases with punishments:

$$\frac{\partial(q^c - q^{fi})}{\partial f} = \frac{\partial q^c}{\partial f} = \frac{1}{2g''(q^c)} < 0. \tag{21}$$

$$\frac{\partial(q^c - q^{fi})}{\partial t} = -\frac{1}{g''(q^c)} > 0. \tag{22}$$

We now show how the intensity of monitoring will be related to monitoring costs, punishments and the price of water. Monitoring is decreasing with the costs of monitoring and increasing with water price and punishment.[20]

$$\frac{\partial m^c}{\partial b} = -\frac{k_2}{b^2} < 0. \tag{23}$$

---

[20] From equation (B4) in the proof of proposition 2 in the appendix, we have

$$\psi'(m_i) = \left\{ \begin{array}{c} \frac{1}{2}\kappa f\left[(q_i - q^{fi}) + (q_j - q^{fi})\right] \\ -\frac{1}{2}f(1-\kappa m_i + \kappa m_j)\frac{\partial(q_j - q^{fi})}{\partial m_i} \end{array} \right\},$$

The partial derivative of $\psi'(m_i)$ with respect to $m_i$ gives

$$\psi^{''}(m_i) = \left\{ \begin{array}{c} \frac{1}{2}\kappa f\left[\frac{\partial}{\partial m_i}(q_i - q^{fi}) + \frac{\partial}{\partial m_i}(q_j - q^{fi})\right] \\ -\frac{1}{2}f(-\kappa)\frac{\partial(q_j - q^{fi})}{\partial m_i} - \frac{1}{2}f(1-\kappa m_i + \kappa m_j)\frac{\partial^2(q_j - q^{f_i})}{\partial m_i^2} \end{array} \right\},$$

$$\text{Where, } \left\{ \begin{array}{l} \frac{\partial(q_i - q^{fi})}{\partial m_i} = -\frac{\kappa f}{2g''(q_i)}; \\ \frac{\partial(q_j - q^{fi})}{\partial m_i} = \frac{\kappa f}{2g''(q_j)}; \\ \text{and} \\ \frac{\partial^2(q_j - q^{fi})}{\partial m_i^2} = \frac{(\kappa f)^2}{4}\left(-\frac{g'''(q_j)}{[g''(q_j)]^3}\right). \end{array} \right.$$

Replacing the above derivatives by their expressions into the expression of $\psi''(m_i)$ and taking into account that the symmetric equilibrium involves that $m_i = m_j = m^c$ and $q_i = q_j = q^c$, yields

$$\psi''(m_i) = \frac{(\kappa f)^2}{4g''(q^c)}\left[1 + \frac{f}{2}\frac{g'''(q^c)}{[g''(q^c)]^2}\right],$$

Given that the cost of monitoring is an increasing and convex function ($\psi''(m_i) > 0$) and $g''(q^c) < 0$, then one gets

$$\left[1 + \frac{f}{2}\frac{g'''(q^c)}{[g''(q^c)]^2}\right] < 0,$$

which implies that $g'''(q^c) < 0$ :

$$g'''(q^c) < -\frac{2}{f}[g''(q^c)]^2 < 0$$

And hence the partial derivative of monitoring with respect to the punishment rate is positive:

$$\frac{\partial m^c}{\partial f} = \frac{1}{b}\left[(q^c - q^{fi}) + \frac{f^2}{8}\frac{g'''(q^c)}{[g''(q^c)]^3}\right] > 0.$$

$$\frac{\partial m^c}{\partial p} = \phi'(k_2)\left(\frac{-f}{g''(q^c)}\right) > 0. \tag{24}$$

$$\frac{\partial m^c}{\partial f} = \frac{1}{b}\left[\left(q^c - q^{fi}\right) + \frac{f^2}{8}\frac{g'''(q^c)}{[g''(q^c)]^3}\right] > 0. \tag{25}$$

A higher water price increases the incentives for theft and the punishment burden that would be incurred by a member who was the only one to be caught, inducing farmers to perform more monitoring to shift the cooperative fine on the others.

In the range of non stringent punishments, there is a higher scope for theft, and an increase in the punishment rate renders the punishment burden for a given level of theft high, and also the total punishment that would be incurred by a member who was the only one to be caught. This would increase the farmers' incentives to compete more in monitoring to shift the cooperative fine on the others. The results above suggest that the distributional effect of peer monitoring is very likely to always dominate the incentive effect.

It is interesting to compare the equilibrium monitoring effort to the socially optimal one. We compare equilibrium outcomes to those that would occur in a second-best problem faced by the WA as a social planner who can decide about monitoring levels of farmers but not their water use choices once monitoring decisions have been made. In addition, we assume that the WA cannot affect the farmers' incentives to steal water for given monitoring levels. In particular, the WA cannot ensure that farmers do not steal the resource. The WA chooses a monitoring level, $m^*$ which maximizes the social welfare function

$$\max_{m \geq 0} 2[g(q^c) - (c+\delta)q^c - \psi(m)]. \tag{26}$$

It is socially optimal not to monitor in cooperatives governed by these rules ($m^* = 0$) whenever monitoring is costly. Farmers over-monitor in equilibrium, $m^c > m^*$, because of their *rent seeking* behavior which results from the dominance of the distributional effect of peer monitoring over the incentive effect.

## 5.4 Endogenous punishment

Here the basic analysis is extended to allow for endogenous punishment, where the punishment rate $f$ is collectively chosen by cooperative members at an initial contracting stage. Inflicting punishment $f$ is costly, we denote by $\varphi(f)$ the associated cost that can be either pecuniary or manifest in nature when there is deterioration of social relations from inflicting punishment on members of a close-knit society. $\varphi(f)$ is increasing and strongly convex (i.e., $\varphi'''(f) > 0$ in addition to $\varphi''(f) > 0$), and satisfies $\varphi(0) = 0$. The strong convexity of $\varphi$ is driven by the increased complexity and difficulty of enforcing more and more stringent punishments on relatives, neighbors and friends.

Cooperative members choose the punishment rate $f^c$ that maximizes an objective function defined as the sum of the cooperative members surpluses, $2\left[g(q^c) - cq^c - pq^{fi} - f\alpha^c - \frac{1}{2}b(m^c)^2\right] - \varphi(f)$ and the WA's surplus, which is equal to its revenue from water proceeds, $2pq^{fi}$ from which is deducted the cost of proving water to the cooperative area, $2\delta q^c$.

$$\max_{f} W^c(f) = 2\left[g(q^c) - (c+\delta)q^c - f\alpha^c - \frac{1}{2}b(m^c)^2\right] - \varphi(f). \tag{27}$$

This has a first-order condition[21]:

$$f^c: -\frac{1}{g''(q^c)}\left(\gamma + \frac{1}{2}f\right) - 2(q^c - q^{fi}) - 2\frac{k_2}{b}\left\{\left[(q^c - q^{fi}) + \frac{f^2}{8}\frac{g'''(q^c)}{[g''(q^c)]^3}\right]\right\} = \varphi'(f) \tag{28}$$

which is also sufficient[22] to identify a global maximum.

From this condition one can show that the punishment level is decreasing with monitoring costs. Totally differentiating the first order condition with respect to $f$ and $b$ and rearranging yields:

$$\frac{\partial f^c}{\partial b} = \frac{2k_2}{Gb^2}\left\{\left[(q^c - q^{fi}) + \frac{f^2}{8}\frac{g'''(q^c)}{[g''(q^c)]^3}\right]\right\} < 0. \tag{29}$$

where[23] $G = \frac{d^2W^c}{df^2} < 0$. This result confirms that the two instruments, monitoring and punishment *complement* each other, as an increase in the cost of one reduces the level of the other.

## 6. Cooperative size

The analysis thus far has remained restricted to the two-farmer cooperative. However, in practice, most cooperatives irrigating from aquifers involve up to as many as 40 farmers, and most cooperatives using surface water involve more than 100 farmers [20]. We investigate here the optimal cooperative size, where the basic set-up is extended from the two-farmer cooperative to the n-farmer one.[24]

We characterize the symmetric subgame perfect equilibrium $(q_n^c, m_n^c)$ for the relevant range of non stringent punishments, $p < f < np$ (assuming that the second-order condition for a maximum holds[25]):

$$q_n^c : g'(q) = c + \frac{1}{n}f, \tag{30}$$

---

[21] Differentiating the cooperative welfare function with respect to $f$ yields

$$\frac{dW^c}{df} = 2\left\{[g'(q^c) - (c+\delta)]\frac{\partial q^c}{\partial f} - \alpha^c - f\frac{\partial \alpha^c}{\partial f} - 2\frac{1}{2}bm^c\frac{\partial m^c}{\partial f}\right\} - \varphi'(f) = 0.$$

Taking into account that $q^{fi}$ does not depend on $f$, we then have $\frac{\partial \alpha^c}{\partial f} = \frac{\partial q^c}{\partial f} = \frac{1}{2g''(q^c)}$. Replacing $\frac{\partial \alpha^c}{\partial f}$ and $\frac{\partial m^c}{\partial f}$ by their expressions given respectively by equations (21) and (25) in the above expression, yields the first-order condition given by equation (28) in the text.

[22] This follows from the strong convexity of the cost of inflicting punishment on cooperative members, ensuring therefore the concavity of the objective function $W^c(f)$.

[23] The negative sign of the second-order derivative follows from the concavity of function $W^c$.

[24] Unlike the two-farmer case where peer monitoring is necessarily mutual, the n-farmer case opens the scope for various kinds of monitoring structures. We will however, momentarily focus on the "mutual monitoring" structure, whereby each farmer monitors all her peers.

[25] It is quite difficult to derive the second-order condition for this problem because the first-order conditions account for the following highly complicated term

$$\phi_n(\kappa m) = (1-\kappa m)^{(n-2)}\sum_{k=1}^{n-1}(1-\kappa m)^{(n-1)(k-1)}.$$

$$m_n^c : k_n \Phi_n(\kappa m) = (n-1)\psi'[(n-1)m]. \tag{31}$$

Where

$$\begin{cases} k_n = f\left[\left(q_n^c - q^{fi}\right) - \frac{1}{n^2} \frac{f}{g''\left(q_n^c\right)}\right] > 0 \\ \text{and} \\ \Phi_n(\kappa m) = \left\{ (1-\kappa m)^{(n-2)(n-1)} \sum_{k=1}^{n-1} (1-\kappa m)^{(k-1)} \right\} \\ = \left\{ \frac{1}{\kappa m} (1-\kappa m)^{(n-2)(n-1)} \left[1 - (1-\kappa m)^{n-1}\right] \right\} \end{cases}.$$

From the necessary conditions one can see that the farmer withdraws more water as the cooperative becomes larger:

$$\frac{\partial q_n^c}{\partial n} = -\frac{1}{n^2} \frac{f}{g''\left(q_n^c\right)} > 0. \tag{32}$$

meaning that larger groups increase the incentives for theft. However, it is not clear whether the equilibrium monitoring level tends to increase or decrease with the cooperative size. The intuition suggests that the group size affects the incentive problem in two ways. A larger group discourages monitoring, as the evidence about the farmer's theft could be established when she is detected stealing by at least one of her peers. Monitoring all together might hence become useless, as the same outcome could be achieved with a smaller number of farmers, avoiding thereby the useless duplication of monitoring. This free-riding problem reduces the farmers' incentives for monitoring. On the other hand, a larger group may increase the total amount of water stolen in the cooperative, increasing therefore the maximum punishment that would be incurred by a member who was the only one to be caught. This would rather increase the farmer's incentives to monitor more intensively to catch the other members stealing, which may reduce her expected share from the total fine. This *rent-increasing effect* will thus counteract the above *free-riding effect* by encouraging more intense monitoring as the cooperative becomes larger.

It is very difficult to derive an analytical expression of monitoring level in equilibrium, as monitoring is implicitly given by (31). In order to get some insights, we will proceed in the remainder of this section to the following simplification: we restrict attention to sufficiently small values of $\kappa$, which implies that all terms in $\kappa^n$ for $n \geq 2$, become of the second-order and can thereby be dropped from our calculations. Consequently (31) reduces to[26]

$$m_n^c : (n-1)[\psi'(n-1)m] \simeq k_n(n-1)[1-(n-1)(n-2)\kappa m]. \tag{33}$$

To obtain explicit solutions where possible we assume that monitoring costs take the quadratic form $\psi(m) = \frac{1}{2}bm^2$ where $b > 0$. Rewriting (33) yields the approximated equilibrium monitoring effort

$$m_n^c \simeq \frac{k_n}{(n-1)[b + k_n \kappa (n-2)]}, \tag{34}$$

[26] This simplification involves no major loss of insights, as equation (33) captures the main qualitative aspects of the solution to the cooperative model.

Monitoring decreases with monitoring costs which is straightforward

$$\frac{\partial m_n^c}{\partial b} = -\frac{k_n}{(n-1)[b + k_n\kappa(n-2)]^2} < 0, \tag{35}$$

As for the impact of the cooperative size on monitoring it is given by

$$\frac{\partial m_n^c}{\partial n} \simeq \theta \frac{\partial k_n}{\partial n}, \tag{36}$$

Where $\theta = \frac{b}{(n-1)[b+k_n\kappa(n-2)]^2} > 0$, which means that the signs of the two partial derivatives$\frac{\partial m_n^c}{\partial n}$ and $\frac{\partial k_n}{\partial n}$ are the same. $\frac{\partial k_n}{\partial n}$ is equal to

$$\frac{\partial k_n}{\partial n} = f\left\{\left(1 + \frac{g'''(q_n^c)}{n^2[g''(q_n^c)]^2}\right)\frac{\partial q_n^c}{\partial n} + \frac{2f}{ng''(q_n^c)}\right\}, \tag{37}$$

Plugging the expression of $\frac{\partial q_n^c}{\partial n} \partial n$ given by (32) into (37) yields

$$\frac{\partial k_n}{\partial n} = \frac{f^2}{ng''(q_n^c)}\left\{\left(2 - \frac{1}{n}\right) - \frac{g'''(q^c)}{n^3[g''(q_n^c)]^2}\right\}. \tag{38}$$

which sign is ambiguous because the term in the bracket parenthesis has an ambiguous sign (the terms $\left(2 - \frac{1}{n}\right)$ and $\frac{g^{(3)}(q^c)}{n^3[g''(q_n^c)]^2}$ are both strictly positive and it is not clear which term overcomes the other). Because of the analytical complexity of the problem, we will address this issue via a numerical example - the example is:

- The production function is $g(q) = \sqrt{q}$;
- The per-unit private cost and price of water are $c = p = 0.2$;
- The transaction costs related to monitoring take two different values $b = 3$ and $b = 10$.

Simulations suggest that the shape and the value of monitoring, $m^c(n)$ as a function of the cooperative size considerably changes when monitoring costs vary. When $b = 3$, the monitoring function gradually decreases as the cooperative becomes larger, i. e., for $n \geq 3$. This means that the *free riding* effect tends to always dominate for small monitoring costs. Whereas, for $b = 10$, monitoring levels become smaller and, the function increases for small values of $n$ and starting from $n = 4$ it gradually declines. This implies that for large monitoring costs the *rent seeking* effect might come into play. Simulation results suggest the existence of a monitoring cost, $\bar{b}$ such that:

- For any $b < \bar{b}$, the equilibrium monitoring $m^c(n)$ is decreasing with the cooperative size and;
- For any $b > \bar{b}$, the equilibrium monitoring $m^c(n)$ increases up to some level $\tilde{n}$ and then gradually declines.

We now explore the optimal cooperative size. Farmers may seek a group size $n_{\max}$ that maximizes the *average*[27] cooperative welfare function $W_A^c(n)$

$$n_{\max} \in \arg\max_{n \geq 2} W_A^c(n) = g(q_n^c) - (c+\delta)q_n^c - \psi(m) \tag{39}$$

The first-order condition for an interior solution (assuming that the second-order condition holds) is given by:

$$[g'(q_n^c) - (c+\delta)]\left(\frac{\partial q^{c_n}}{\partial n}\right) - \psi'(m_n^c)\left(\frac{\partial m_n^c}{\partial n}\right) = 0, \tag{40}$$

The (first-order) change in social welfare attributable to a marginal entrant is composed of two terms. The first term implies that the new entrant causes every member to better free ride on her peers and thus to contract her monitoring effort. This would increase the opportunities for theft for everyone. This *stealing* effect causes a reduction in social welfare of

$$[g'(q_n^c) - (c+\delta)]\left(\frac{\partial q_n^c}{\partial n}\right) < 0, \tag{41}$$

On the positive side, free riding results in monitoring cost savings. This *cost savings* effect brings about an increase in social welfare of

$$-\psi'(m_n^c)\left(\frac{\partial m_n^c}{\partial n}\right) > 0. \tag{42}$$

The optimal cooperative size, $n_{\max}$ equates the social marginal benefit stemming from monitoring cost savings to the social marginal losses caused by a higher occurrence of theft. The net benefits of peer monitoring are maximized when the cooperative size is neither too small (due to the "*monitoring cost savings*" effect) nor too large (due to the "*stealing*" effect).

The effect of the group size on the cooperative welfare is found to be analytically complicated, that is why we use a numerical example to shed light on the intensity of *stealing* and *cost savings* effects when one varies monitoring costs. Simulations are performed for the same production function and the same parameter-values considered above, to which we add the value of the external cost of water $\delta = 3$. Simulation results suggest that for $b = 3$, the welfare function is maximized for $n_{\max} = 4$, while in the other case it is gradually decreasing. This means that the *stealing* effect dominates almost everywhere and the best policy of the WA is to implement small and medium cooperatives.

Finally, simulations suggest that monitoring costs reduce the cooperative size. The intuition is that higher monitoring costs make it more difficult to monitor, which gives more opportunities for theft, requiring smaller cooperatives to compensate.

## 7. Extensions

### 7.1 Collusion

The cooperative model described up until now corresponds to a non-cooperative game. Each cooperative member is out to maximize her expected payoff, and makes

[27] The rational for the choice of the average social welfare function rather than the absolute one is that for the latter the group size effect might always dominate and the function is very likely to always increase with the cooperative size.

her decisions about monitoring and water use independently of the other members. What happens if we relax this assumption and consider possibilities of coordinated actions about monitoring? A natural model is to consider what happens if the two cooperative members choose their monitoring efforts in order to maximize joint payoffs, $\left[U_i(q_i, m_i) + U_j\left(q_j, m_j\right)\right]$. The collusive outcome is given by corollary 3:

**COROLLARY 3**: *The collusive monitoring efforts are*

$$m_i = m_j = 0. \tag{43}$$

**Proof:** See the appendix.

The collusive monitoring effort is efficient. In the absence of collusion, cooperative members compete on monitoring because of their *rent-seeking* behavior, though monitoring is useless, as they will always split equally the cooperative fine between themselves.[28] Collusion is thus welfare enhancing, as it yields the same outcome and saves in monitoring costs.

## 7.2 Monitoring structures

Although in practice the mutual monitoring structure - whereby each farmer in the group is being simultaneously monitored by all of her peers - is commonly observed [15], other monitoring structures deserve consideration. An interesting departure from the mutual structure is the "*localized monitoring*" in which every farmer monitors (and is monitored) by only one of her closest neighbors, say each farmer monitors her left neighbor and is monitored by her right neighbor. There is a natural argument in favor of monitoring structures of the latter kind: it can potentially avoid duplication[29] of monitoring. As a first and very tentative attempt to explore the issue of the optimal design of peer monitoring structures, we shall compare the mutual monitoring (MM) to the localized monitoring (LM) one with regard to the equilibrium water use and monitoring levels and thereby to the cooperative welfare level. The comparison will be held for a three-farmer cooperative.

### *7.2.1 Mutual monitoring*

Consider a cooperative formed by three farmers $i$, $j$ and $k$ applying *mutual peer monitoring*. We assume that a farmer monitors each of her peers with the same monitoring effort, which means that the cost of monitoring all others members is equal to $\psi(2m_l)$ for $l = i, j, k$..

The subgame perfect equilibrium is the profile $\left(m_l^{c(MM)}, q_l^{c(MM)}\right)_{l=i,j,k}$ of monitoring efforts $m_l^{c(MM)} \geq 0$ and water use levels, $q_l^{c(MM)}$ mapping from the set of monitoring decisions into the set of water use decisions, $q_l^{c(MM)} : [0, +\infty)^3 \to 0, \tilde{Q})$

[28] This finding comes from the symmetric equilibrium which implies that cooperative members perform the same level of monitoring, i.e., $m_i^c = m_j^c = m^c$. It follows that the expected share of farmer $i$ from the total fine is equal to $\frac{1}{2}$ :

$$\begin{aligned} s_i^{\exp} &= \frac{1}{2}\left(1 - \kappa m_i^c + \kappa m_j^c\right) = \frac{1}{2}(1 - \kappa m^c + \kappa m^c) \\ &= \frac{1}{2} \end{aligned}$$

[29] Duplication in the mutual structure obviously takes place when the number of farmers in the cooperative is larger than two.

(where the superscript $^{c(MM)}$ refers to cooperatives characterized by mutual monitoring). In what follows we shall focus on *symmetric* equilibria which imply that

$$q_l^{c(MM)} = q_3^{c(MM)} \text{and} m_l^{c(MM)} = m_3^{c(MM)} \text{for } l = i, j, k.$$

(assuming that the second-order condition for a maximum holds). The solution to this problem is summarized in proposition 3:

**PROPOSITION 3**: *If* $p < f < 3p$, *there exists a unique symmetric subgame perfect equilibrium* $\left(m_3^{c(MM)}, q_3^{c(MM)}\right)$ *satisfying*

$$q_3^{c(MM)} : g'(q) = c + \frac{1}{3}f, \tag{44}$$

and

$$m_3^{c(MM)} : \left\{ \begin{array}{c} -\dfrac{f^2}{9g''\left(q_3^{c(MM)}\right)} \begin{bmatrix} \kappa^4 m^4 + \dfrac{1}{2}\kappa^4 m^3 \\ -2\kappa^3 m^3 - \dfrac{5}{2}\kappa^3 m^2 \\ +3\kappa^3 m \end{bmatrix} \\ -f\left(q_3^{c(MM)} - q^{fi}\right) \begin{bmatrix} \kappa^4 m^3 - 3\kappa^3 m^2 \\ +4\kappa^2 m - 2\kappa \end{bmatrix} \end{array} \right\} = 2\psi'(2m). \tag{45}$$

**Proof**: See the appendix.

### 7.2.2 Localized monitoring

By the same token, consider a three-farmer cooperative in which farmers $i$, $j$ and $k$ apply *localized monitoring* whereby each farmer monitors her left neighbor and is monitored by her right neighbor. We assume that farmer $i$ monitors farmer $j$ with the monitoring level $m_{ij} = m_i$, and farmer $j$ monitors farmer $k$ with the monitoring effort $m_{jk} = m_j$, and farmer $k$ monitors farmer $i$ with the monitoring effort $m_{ki} = m_k$. We characterize the symmetric subgame perfect equilibrium (the superscript $^{c(LM)}$ refers to cooperatives characterized by localized monitoring)

$$q_l^{c(LM)} = q_3^{c(LM)} \text{ and } m_i^{c(LM)} = m_3^{c(LM)} \text{for} l = i, j, k.$$

(assuming that the second-order condition for a maximum holds). The solution to this problem is characterized by proposition 4:

**PROPOSITION 4**: *If* $p < f < 3p$, *there exists a unique symmetric subgame perfect equilibrium* $\left(m_3^{c(LM)}, q_3^{c(LM)}\right)$ *which satisfies*

$$q_3^{c(LM)} : g'(q) = c + \frac{1}{3}f, \tag{46}$$

and

$$m_3^{c(LM)} : \left(-\frac{1}{2}\kappa^2 m + \kappa\right) \begin{bmatrix} -\dfrac{f^2}{9g''\left(q_3^{c(CI)}\right)} \\ +f\left(q_3^{c(CI)} - q^{fi}\right) \end{bmatrix} = \psi'(m). \tag{47}$$

**Proof:** See the appendix.

Comparing the monitoring levels in both settings is not straightforward, however under certain relevant circumstances one can say something. For sufficiently small $\kappa$, the terms $\kappa^n$ for $n \geq 2$, become of the second-order and equations (45) and (47) above reduce respectively to

$$\kappa f\left(q_3^{c(MM)} - q^{fi}\right) \simeq \psi'\left(2m_3^{c(MM)}\right). \tag{48}$$

and

$$\left[-\frac{\kappa f^2}{9g''\left(q_3^{c(LM)}\right)} + \kappa f\left(q_3^{c(LM)} - q^{fi}\right)\right] \simeq \psi'\left(m_3^{c(LM)}\right). \tag{49}$$

A mere comparison of (48) and (49) suggests that the equilibrium localized monitoring is higher than twice the equilibrium mutual monitoring

$$m_3^{c(LM)} > 2m_3^{c(MM)}. \tag{50}$$

This result may sound unlikely, one indeed expects farmers to monitor less in cooperatives characterized by localized monitoring, as this structure avoids the duplication of monitoring. But, the rationale for this result comes from the distributional character of monitoring, as cooperative members compete on monitoring to shift the cooperative fine on the others. In the mutual structure, a farmer may well reduce her expected share from the total fine as the cooperative fine might be shared with more than one member. However, in the localized structure, each farmer monitors only one of her neighbors increasing the risk to bear the whole punishment burden on her own. This acts as an incentive to monitor more to increase the probability of shifting the fine on that neighbor.

An immediate implication of (44), (46) and (50) is that cooperatives using mutual monitoring welfare dominate those using localized monitoring

$$W^{c(MU)}(3) > W^{c(CI)}(3). \tag{51}$$

Where $W^{c(MU)}(3)$ and $W^{c(CI)}(3)$ are welfare levels in the three-farmer cooperatives characterized by mutual and localized monitoring structures, respectively.

## 8. Policy implications

Surface water is not only scarce but also highly uncertain and often of bad quality in particular in arid and semi-arid regions. Groundwater is seen as a unique guarantee (whenever it is well managed) for the long term viability of an agriculture sector crucial for giving those fragile countries a minimum food security.

The principal aim of this research was to design the appropriate institutions and incentives to reduce groundwater over-exploitation. We have proposed various incentive schemes, some based on individual water use and others on aggregate water use. The model has in particular shown that total water use-based incentive schemes may well overcome the individual water use-based scheme (centralized management). Moreover, the comparative evaluation of the two team-based incentive schemes has shown that for a relevant range of non stringent punishments, the first scheme dominates the cooperative management since it restores the full-information outcome. Nevertheless such a scheme may suffer from a strong implementation problem: endowment constraints may render it infeasible. How to overcome this drawback? The solution could be imported from the theory of "*positive*" assortative matching, where farmers could be sorted into various homogenous groups giving the scope to policy makers to choose the team-based scheme to implement for each group. To illustrate this idea, suppose that the WA could sort farmers into two groups, a group of wealthy farmers and a group of all other farmers. How to screen wealthy farmers from others? Wealthy farmers may well be those who invest more in sophisticated agricultural machinery and/or grow high value crops and/or cultivate large plots of land,...etc. For the group of wealthy farmers, the optimal policy of the WA would be to implement the first team-based scheme; And for the other group, participatory management could be implemented through the creation of a collective responsibility rule that induces peer monitoring by members as modeled above.

Simulations in our paper suggested that the cooperative size is weakly decreasing with monitoring, meaning that the best policy of the WA would be to implement non large cooperatives which would reduce the scope for groundwater over-exploitation.

## 9. Potential extensions for further research

This research could be extended along the following directions. Firstly, we know from simulation results that it is socially beneficial to form small and medium cooperatives. But, what about irrigated areas with a large number of farmers? In practice, centralized structures are in charge of running such areas. In addition, a direct management transfer of these areas to cooperative institutions is unlikely to enhance water use efficiency, as the "*stealing effect*" a major source of inefficiency always dominate in large cooperatives. The cooperative management could however be implemented in a slightly different way through the creation of two hierarchies of cooperatives, *primary* and *secondary* cooperatives. Farmers can form several groups of medium and/or small sizes depending upon whether peer monitoring involves low or high costs. These groups would constitute *secondary cooperatives* where each is run by a number of its members, called "cooperative manager". The secondary cooperatives' managers themselves would form what we call a "*primary cooperative*" playing the role of intermediary between secondary cooperatives and the WA. Among each secondary cooperative's managers, some farmers would be chosen to form the primary cooperative's managers and would be in charge of running it. Every cooperative is governed by similar rules as modeled above: members of the same group are related by some collective responsibility rule which would create incentives for two types of peer monitoring, (a.) *peer monitoring-within* applied within each secondary cooperative and (b.) *peer monitoring-between* applied between secondary cooperatives.

Secondly, in our theoretical study we have formalized the cooperative management where members interact only once, however in practice members interact repeatedly. The knowledge that pursuit of their short-term interests can harm their long-term aims by affecting the reaction of others in the future interactions may be a powerful inducement to behavior that displays apparent solidarity with the interests of the group. When cooperative members are homogenous, in particular they are all landlords and moreover the agricultural activity is their main source of income, their interactions can be modeled as an infinitely repeated game.[30] What one can learn from group lending literature is that such cooperatives might not need to have access to an exogenous penalty device since peer sanctions can be accomplished in the dynamic framework and they are self enforcing [31] [17]. It suffices to make the group members jointly liable for the payment of their total amount of water use, e.g., denying them future access to the public source of water supply or rising the price of water for next periods when theft occurs in the current period. In such a management design, a group member can be penalized by other members' shirking (=stealing), in that a member's shirking increases the payment burden of her peers, and thereby negatively affecting their payoffs. In turn, when cooperative members are heterogenous in that some members are landlords and others are rental contract holders and/or tenants, an exogenous penalty technology might be required, e.g., one may think for example of the exclusion of defaulting users from the cooperative or from the community or from certain kinds of input supply facilities.[32] Whereas, being excluded from the cooperative could be perceived as extremely harmful for a land owner, this might not be the case for rental contract holders who interact only for a finite number of periods and the pursuit of their short-term interests induce them to adopt the strategy of "take the money and run away.[33]" However, this may well harm their long-term interests - the loss of reputation may be very costly for these farmers: their exclusion from the cooperative might make it difficult for them to integrate other cooperatives in the same area and/or even in other areas. In short they might be excluded from the community. What might be critical here is the enforcement of such social sanctions in practice, especially how and why a farmer should ever impose a sanction on a "friend" or "relative" who has defaulted. One possible explanation on the face of it is the impossibility to keep the information about strategic defaults secret. An other explanation may be the harmful consequences of foregoing the punishment of defaulters, e.g., the absence of alternative sources of water supply if the cooperative is denied future access to her principal source of water supply.

## 10. Conclusion

This paper has investigated the design of the appropriate institutions and rules to enhance groundwater use efficiency by reducing over-pumping of aquifers. We have proposed various incentive schemes, some based on individual withdrawals which are the farmers' private information and some based on the total water withdrawn by

---

[30] Actually, landlords interact for a finite number of periods but, there is enough uncertainty about the end of their interactions that this can be modeled as an infinite repeated game.

[31] Yeon-Koo Che [17] builds a group lending model where the incentive problem stems from the entrepreneurs' unobservable effort decisions and their liquidity constraints. He shows that when the group members operate their projects repeatedly, the joint liability feature itself makes it credible for members to penalize others through their effort decisions. Under group lending, a member's shirking in her productive effort increases the payment burden of her peers, i.e., a group member can be penalized by other members' shirking.

[32] Armendariz De Aghion [15] reports that social sanctions are observed in practice. For example, in agricultural cooperatives, social sanctions often involve the exclusion of defaulters from privileged access to input supplies and marketing facilities.

[33] At the end of her contract, a rental contract holder may well benefit from water theft without incurring the cost of stealing because she can refuse to pay the punishment since she would leave the cooperative anyway.

all farmers which is publicly observable. In the latter setting, two schemes are proposed. In the first scheme, the WA administers an incentive scheme that does not balance the budget, restoring thereby the full-information water use level. Such scheme works independently of the group size, but it may be infeasible when farmers have endowment constraints. This is why the WA resorts to a second total water use-based incentive scheme by promoting the cooperative behavior. We device cooperative institutions characterized by a joint liability clause that induces peer monitoring by members. We show that higher monitoring costs and larger cooperatives entail more theft and higher punishment levels reduce it. We also show how the cooperative membership and punishments are determined endogenously by constraints on monitoring. Higher monitoring costs increase punishment levels and reduce the size of the cooperative. The basic analysis is then extended to allow first for collusion in monitoring between cooperative members, and show that the collusive monitoring effort is efficient. Secondly, we explore a different monitoring structure "localized monitoring" and compared it with the mutual monitoring structure which is commonly observed in practice. Finally, we use the theoretical results to derive some useful policy recommendations that could help decision makers to implement the right policies to alleviate groundwater over-exploitation.

Overall, these results provide strong confirmation of the ability of well designed incentives and institutions to reduce groundwater over-exploitation, and that constraints on monitoring costs affect institutional design.

## Appendix

The details of mathematical demonstrations are available from the authors upon request.

## Classification

JEL classification: Q13; Q15; Q25; R48

## Author details

Wided Mattoussi[1]*, Mohamed Salah Matoussi[2] and Foued Mattoussi[3]

1 Ecole Supérieure des Sciences Economiques et Commerciales de Tunis (University of Tunis) and Laboratoire de Recherche en Economie Quantitative du Développement (LAREQUAD), Tunis, Tunisia

2 Faculté des Sciences Economiques et de Gestion de Tunis (University of Tunis El Manar) and Laboratoire de Recherche en Economie Quantitative du Développement (LAREQUAD), Tunis, Tunisia

3 Faculté des Sciences Juridiques, Economiques et de Gestion de Jendouba (University of Jendouba) and Laboratoire de Recherche en Economie Quantitative du Développement (LAREQUAD), Jendouba, Tunisia

*Address all correspondence to: wided.mattousse@gmail.com

## References

[1] Falkenmark, M. 2005. Water usability degradation – economist wisdom or societal madness?. *Water International* 30(2):136-146.

[2] Igor, S. Z. and G.E. Lorne. 2004; Groundwater Resources of the World and their Use. Published by the United Nations Educational, Scientific and Cultural Organization, place de Fontenoy, 75352 Paris 07; France.

[3] Schmoll, O.; Howard, G.; Chilton, J.; Chorus, I. & World Health Organization, Water, Sanitation and Health Team. 2006. Protecting groundwater for health: managing the quality of drinking-water sources. 1st ed.; IWA Publishing: Alliance House, 12 Caxton Street, London SW1H 0QS, UK, 69

[4] Foster, T.; Brozovi, C. S. 2017. The buffer value of groundwater when well yield is limited. *Journal of Hydrology* 547: 638-649.

[5] Katic, P. and R.Q. Grafton. 2011. Optimal Groundwater Extraction under Uncertainty: Resilience versus Economic Payoffs. *Journal of Hydrology* 406(3-4): 215-224.

[6] Tsur, Y. and G. Tomasi. 1991. The buffer value of groundwater with stochastic surface water supplies. Journal of Environmental Economics and Management 21:201-224.

[7] Smith, M., Cross, K., Paden, M. and P. Spring. Laban. 2016: Managing groundwater sustainably. Publisher: IUCN, Global Water Programme Rue Mauverney 28 1196 Gland, Switzerland.

[8] Koundouri, P. 2004. Current Issues in the Economics of Groundwater Resource Management. *Journal of Economic Surveys* 18(5):703-740.

[9] Giordana, G. A. and M. Montginoul. 2006. Policy instruments to fight against seawater intrusion in coastal aquifers: an overview. *Vie et Milieu - Life and Environment* 56(4):287-294.

[10] Howe, C. W. 2002. Policy issues and institutional impediments in the management of groundwater: lessons from case studies. *Environment and Development Economics* 7(4):625-641.

[11] Mattoussi, W. and P. Seabright. 2014. Cooperation against theft: a test of incentives for water management in Tunisia. *American Journal of Agricultural Economics* 96(1):124-153.

[12] Lenouvel V.; Montginoul, M. and S. Thoyer. 2011. From a blind truncheon to a one-eyed stick: testing in the lab an optional target-based mechanism adapted to groundwater withdrawals. Paper presented at the 18th annual conference of European Association of Environmental and Resource Economists, Rome.

[13] Johansson, R.C, and Y. Tsur, L.R. Terry, R. Doukkalid and A. Dinar. 2002. Pricing irrigation water: a review of theory and practice. *Water Policy* 4: 173–199.

[14] Holmström, B. 1982. Moral Hazard in Teams. *Bell Journal of Economics* 13: 324-340.

[15] Armanderiz De Aghion, B. 1999. On the design of a credit agreement with peer monitoring. *Journal of Development Economics* 60:79-104.

[16] Besley, T.; Coate, S. 1995. Group lending, repayment incentives and social collateral. *Journal of development economics* 46:1-18.

[17] Che, Y.K. 2002. Joint liability and peer monitoring under group lending. Unpublished, University of Wisconsin-Madison.

[18] Conning, J. 2005. Monitoring by delegates or by peers? joint liability loans under moral hazard" Unpublished, Hunter college and the Graduate Center.

[19] Stiglitz, J.E. 1990. Peer monitoring and credit markets. The world bank economic review 4(3):351-366. 31.

[20] Ghatak, M.; Guinnane, T.W. 1999. The economics of lending with joint liability: Theory and practice. *Journal of development economics* 60:195-228.

[21] Varian, H.R. 1990. Monitoring agents with other agents. Journal of Institutional and Theoretical Economics 146:153-74.

[22] Montginoul, M.; Rinaudo, J. D.; Brozovic¥, N. and G. Donoso. 2016. Controlling Groundwater Exploitation Through Economic Instruments: Current Practices, Challenges and Innovative Approaches. A.J. Jakeman et al. (eds.), Integrated Groundwater Management.

[23] Casari, M.; Plott, C.R. 2003. Decentralized management of common property resources: experiments with a centuries-old institution. *Journal of Economic Behavior and Organizations* 51 (2):217–247.

[24] Segerson, K. 1988. Uncertainty and incentives for nonpoint pollution control. Journal of Environmental Economics and Management 15(1): 87-98

[25] Xepapadeas, A.P. 1991. Environmental policy under imperfect information: incentives and moral hazard. Journal of Environmental Economics and Management 20:113-126.

[26] Miceli, T. and K. Segerson. 1991. Joint liability in torts and infra-marginal efficiency. International Revue of Law and Economics 11:235-249.

[27] Karp, L. 2005. Non-point Source Pollution Taxes and Excessive Burden. *Environmental and Resource Economics* 31 (2):229–251.

[28] Millock, K. and F. Salanié. 2005. Nonpoint source pollution when polluters might cooperate. Topics in Economic Analysis & Policy 5(1), Article 12.

[29] Litchenberg, E. 1992. Alternative Approaches to Pesticides Regulation. *Northeastern Journal of Agricultural and Resource Economics* 21:83–92.

[30] Ribaudo, M.; Horan, R.D. and M.E. Smith. 1999. Economics of Water Quality Protection From Nonpoint Sources: Theory and Practice. Resource Economics Division, Economic Research Service, U.S. Department of Agriculture, Agricultural Economic Report No. 782.

[31] Shortle, J.S. and R. D. Horan2001. The economics of nonpoint pollution control. Journal of Economic Surveys 15 (3):255-289.

[32] Giordana, G. 2007. Experimentation d'instruments fiscaux pour gérer l'exploitataion des aquifères côtiers. PHD. Thesis in Economics, University of Montpellier, France.

[33] Figureau, A.G.; Montginoul, M.; Rinaudo, J.D. 2015. Policy instruments for decentralized management of agricultural groundwater abstraction: a participatory evaluation. *Ecological Economics* 119:147–157.

Chapter 6

# Nanoparticles in Wastewater Treatment

*Farid Alizad Oghyanous*

## Abstract

Water plays a crucial role in every animate life. There are a multitude of problems that can be occurred without water; thereafter, mankind's lives can be extinct. Several solutions should be implemented in order to protect water supplies and to treat water used in industries. Among solutions, wastewater treatment is sounded economical and convenient way to overcome water scarcity. Physical, chemical, biological, and mixed treatment systems provide ample opportunity to use water over and over again. However, by using nanotechnology in these systems wastewater treatment can reach much more quality and overcome their drawbacks. Nanomembranes in MBR technology is one of most appropriate treatment technologies that have such potential to postpone water shortage until several years.

**Keywords:** water scarcity, wastewater treatment, water reuse, nanoparticles, nano-materials, nanocomposite membranes

## 1. Introduction

Let us imagine a world without water. When you get up in the morning, the first place you usually go is the lavatory and use water to feel fresh and to get rid of the tiredness of the night. But, what will happen if water is not flowing from the faucet or pipes? Perhaps, on the first day, it would impair your ability to focus on your work besides thinking about water would deter you from doing even your daily chores and errands. Indeed, lack of water can affect all people's life adversely like making them sick or even causing them to pass away in just a few days.

Hence, why water is so important in mankind's life? Why should all people do their best to protect water supplies? The imagination of the world without water is terrible; therefore, what will happen if individuals do not have water in reality? There is an enormous amount of water all over the Earth, so why we have to hear the news of water scarcity every day on news? Is it possible for people to reuse water that was used before? What are the best ways of treating water and wastewater in order to use them over and over again? We are going to discuss all of these questions and answer them whereby our aim is that people try to protect WATER from being polluted.

Every organism, regardless of micro-organisms and macro-organisms, needs water to continue living. Some micro-organisms could live without oxygen, however, there are no animated that could live without water. So, water is absolutely more important than oxygen by which human beings and marine animals are breathing. Indeed, their lives rely on water due to the fact that their bodies comprise a lot of water and they need to consume water to gain energy and to metabolism activity.

Insufficient water causes the human body's systems to change. Almost two billion people do not have access to basic water services in public health care facilities around the world. Hand washing with soap has a proven health benefit, but in reality, one out of six health facilities worldwide lacks functioning hand washing facilities either at clinics or toilets. It is widely believed that training in basic water, sanitation, and hygiene is essential for making a healthy learning environment. Without that, cells of the body shrink. Mankind's brain will tell their bodies to urinate less. A kidney's ability to eliminate waste from the blood depends on adequate water levels. People's kidneys cannot function properly without adequate water. Without enough water, the kidneys have to work harder and wear down tissues. Water is essential to the health of the kidneys and other organs in the body. Without adequate water intake, these organs and others may not function properly. A lack of water will result only in a matter of days before a person dies, so people need more than enough water in their bodies. Keep in mind that exercise, hot temperatures, and illness can increase the need for water for staying healthy. Generally, thirst will guide individuals when it comes to how much water they should drink [1].

It can be inferred from the aforementioned determinants of water that without water humans will be extinct. Even though water is utilized by living organisms and so necessary for them, the main water using pertains to industries and agricultural activities around the world. Living organisms' share of use of water in comparison with factories and agriculture activities is negligible. Various kinds of factories and industries use water as a raw material to produce beverages, to manufacture cars, to separate crude oil, to irrigate their farms, etc. Under such circumstances, water being polluted since it is used in different kinds of processes and related to its process usage, its contaminations could be various. If the world has unlimited water sources, it is not even acceptable to use water in the factories and pour them somewhere else without treatment because its contamination would spread pathogen microbes. Oceans have plenty of water, however, it is salty water and it is needed ridiculously expensive processes to eliminate its salt so that it can be used. Therefore, our world does not have sweet water, so industries have to treat water used in their processes. Treating wastewater is much more convenient than seawater. Treating and reusing wastewater provides people with a water cycle and it postpones water scarcity until several years later. So, not only do governments enact serious laws in order to oblige industries to treat their used water in a very appropriate way, each individual had better do their responsibilities not to use water more than their needs. Eventually, Water scarcity and the increasing demand for clean water make treating and recycling wastewater inevitable in the Twenty-First Century.

## 2. Wastewater treatment methods

Wastewater treatment methods can be categorized into three main areas: (1) physical, (2) chemical, and (3) biological. Wastewater treatment using physical processes is primarily dealing with solid–liquid separations, in which filtration plays a major role. Conventional and nonconventional filtration techniques are divided into two general categories. Water treatment applications rely on this technology. The treatment process is only one unit of a conventional water treatment scheme, in which there is a wide range of equipment and technology options to choose from depending on the ultimate goal of treatment. It is important to understand the role of filtration in water purification in comparison to other technologies as well as the objective of different unit processes. This economic process can remove suspended solids of the wastewater and also in some cases like using membranes it can eliminate micro-organisms of wastewater. Nevertheless, it is not able to decrease organic contaminations and heavy metals of the wastewater alone which are so harmful either in reusing at industries or domestics. Membrane filtration is one of the dominant examples of this process whose structure not only can be modified by using some novel technology like nanoparticles but can be exploited with other methods of treatment easily.

Chemical methods of treatment rely on chemical interactions between the contaminants and the operator of the chemical apply and provide assistance in either removing contaminants entirely from water or neutralizing harmful effects associated with contaminants. It is possible to apply chemical treatment methods standalone and also as a part of a treatment process that involves physical processes. Organics of the wastewater will be removed by utilizing this expensive method; however, it will enter some new compounds into the wastewater some of which might be harmful. For instance, adsorption by activated carbon is commonly used in industries and domestic treatments in order to remove turbidity and the scent of water without any side effects.

The biological treatment of wastewater has an apparent simplicity because it relies on natural processes to help with the decomposition of organic compounds, but it is actually complex, not fully understood, and taking place at the intersection of biology and biochemistry. Wastewater contains organic matter, including garbage, organic wastes, partially digested foods, heavy metals, and toxins. Biological treatments tend to rely on bacteria, nematodes, and other small organisms to break down organic matter. Biological treatment can be used worldwide because it is flexible, economic, and environmentally friendly. Many mechanical or chemical processes cannot match the effectiveness or efficiency of biological treatments. The conventional activated sludge (CAS) process is a good illustration of this. These systems typically include an aeration tank that acts as a biological degradation agent and a secondary clarifier for separating sludge from treated wastewater [2].

Wastewater treatment methods can incorporate any combination of these three technology groups, or select ones, depending on the treatment objectives. According to the reusing standards, these methods can be combined with one another and it is quite obvious that if anyone wants to reach high quality of treated water, they should combine all three methods in order to reach the water which even can be drunk. Totally, the main factors affecting the way of treatment methods choosing are the characteristics of the wastewater and the quality needed for reusing. Depending on the manufacturing process, industrial wastewater can contain specific organic constituents, high salinity, heavy metals, acids with high pH, and inorganic particles with high turbidity. In this regard, a wastewater treatment plant, either industrial, municipal, or drilling water treatment, is not much different from, say, a rubber factory or an oil refinery. In most industrial and agricultural activities, high-quality clean water is crucial; however, treating water to high standards is expensive; and droughts, contamination of water resources, and competing demands inevitably limit the availability of water for agriculture and oil and gas extraction, complicating food production and energy production. The successful treatment of poor quality water must not only produce water of desirable quality but also protect downstream processes. In industrial wastewater treatment, multiple sequential treatment steps are often needed. This is called a "treatment train." The cost of distilling contaminated water is prohibitive and not all industrial waters require this level of treatment. While distillation can remove all contaminants, it may not always be feasible. Moreover, certain technologies have inherent limitations, particularly at high water recovery and salinity. Membranes are limited by osmotic and hydraulic pressures, as well as mineral precipitation. Adsorbers are limited to certain compounds having specific functional groups. Specifically, unit processes for removing contaminants from a range of chemical groups are needed, as is the ability to convert a variety of industrial waste streams into high-value materials and energy. Dedicated wastewater treatment plants often pre-treat industrial wastewater on-site. In terms of sustainable economic development and the environment, new technologies for the treatment of impaired and unconventional water are crucial. By using nanotechnology, waste streams can be treated, allowing water to be reused and energy to be recovered, as well as highly valuable materials [3–5].

## 3. Nanotechnology in wastewater treatment

Conventional water treatment is not always very effective at removing contaminants such as metals and micro-organisms. The formation of disinfection byproducts (DPBs), which can harm human health, is another problem. The DPBs are formed when chemical disinfectants react with organic matter and inorganic ions in the water. The removal of metals, microbes, and oil from contaminated water has been studied using nanomaterials in some studies. Nanomaterials have been used as an alternative to remove contaminants. In nanoscience, phenomena are studied at the nanometer scale. Nanotechnology involves materials with at least one component whose dimension is less than 100 nm [6]. These materials differ greatly from conventional materials, in terms of mechanical, electrical, optical, and magnetic properties, due to their nanoscale size. The nanomaterials' small size and large surface area make them highly adsorbent and reactive. In addition, nanomaterials have been reported to be highly mobile in solution. Heavy metals, organic pollutants, inorganic anions, and bacteria have all been reported to be removed using various types of nanomaterials. Many nanomaterials have been extensively investigated for their potential applications in water and wastewater treatment, including zero-valent metal nanoparticles, metal oxide nanoparticles, carbon nanotubes, and nanocomposites.

### 3.1 Zero-valent metal nanoparticles

#### *3.1.1 Silver nanoparticles*

Antibacterial properties of silver nanoparticles have been attributed to their high toxicity to microorganisms, including bacteria, viruses, and fungi. Since silver nanoparticles are good antimicrobial agents, they are widely used to disinfect water. There is no clear understanding of how Ag NPs have antimicrobial effects, and the mechanisms remain unclear. The adherence of Ag nanoparticles to bacterial cell walls and subsequent penetration, which caused structural changes within the membrane and thus increased its permeability, has been hypothesized. Moreover, when Ag NPs contact bacteria, free radicals are produced. These free radicals damage cell membranes, causing the death of cells. DNA contains abundant sulfur and phosphorus elements, which also cause the death of cells. More importantly, the NPs when broken down will release $Ag^{+}$ ions, which will interact with enzyme thiol groups, inactivate them, and hinder normal functions of the cell [7–9].

#### *3.1.2 Iron nanoparticles*

Recently, zero-valent metal nanoparticles, such as Fe, Zn, Al, and Ni, have been gaining wide research interest in water treatment. Nanozero-valent aluminum is thermodynamically unstable in water due to its high reductive abilities, which leads to formation of oxides/hydroxides on the surface, interfering entirely with electron transfer from the metal surface to contaminants. The standard reduction potential of Ni is less negative than that of Fe, indicating lower reduction ability, while nano-zero-valent Fe or Zn has a moderate standard reduction potential and are ideal reducing agents relative to most redox-labile contaminants. While Fe has weaker reduction abilities, it has many effects on water pollution and is an excellent adsorbent, precipitates and oxidizes (if oxygen is present), and is relatively inexpensive. So far, zerovalent iron nanoparticles have been subjected to the most extensive study among zerovalent metal nanoparticles [10].

### 3.2 Metal oxides nanoparticles

#### *3.2.1 $TiO_2$ nanoparticles*

It has been proven that photocatalytic degradation technology can successfully be used in the treatment of water and wastewater by oxidizing contaminants into low molecular weight intermediate products, which eventually turn into $CO_2$, $H_2O$, and anions such as $NO_3^-$, $PO_3^-$, and $Cl^-$ for reuse. Metal oxides and sulfide semiconductors are the most common photocatalysts. Of them, $TiO_2$ has been investigated most intensively in recent decades due to its high photocatalytic activity, reasonable price, photostability, and chemical and biological stability [11]. Besides, Due to its low cost, toxic free property, chemical stability, and easy availability on earth, titanium dioxide ($TiO_2$) is one of the best photocatalysts existed on the earth. Anatase, rutile, and brookite are among the three natural states of $TiO_2$. Until today, Anatase is considered a good material for nanophotocatalysis [12]. In short, its running process described as below: a semiconductor like $TiO_2$ absorbs light that is greater or equal to its band gap width, carrying electron–hole pairs ($e^-$-$h^+$). By separating the charge further, the electrons and holes may travel to the catalyst surface, where they are combined with the sorbed species to produce the redox reactions. The hydroxyl radicals are generated when $h^+_{vb}$ react with water (surface-bound) and the radical anion (superoxide radicals) when $e^-_{cb}$ selected by oxygen, as shown below in **Figure 1** [13].

#### *3.2.2 ZnO nanoparticles*

Apart from $TiO_2$ nanoparticles, ZnO nanoparticles have emerged as a valuable photocatalytic candidate in water and wastewater treatment due to their unique properties, including high oxidation capacity and good photocatalytic property. In addition to being environment-friendly, ZnO NPs are also compatible with organisms, which make them ideal for sewage treatment. Their photocatalytic capacity is similar to that of $TiO_2$ NPs because their band gap energies are similar. In contrast to TiO2 NPs, ZnO NPs are more affordable [14]. ZnO nanoparticles can

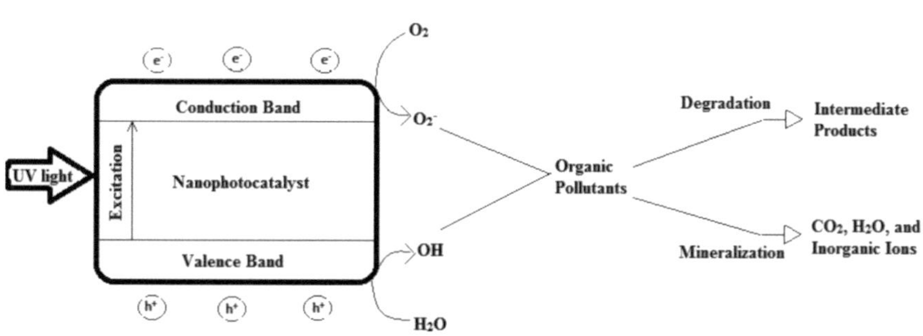


**Figure 1.**
*Nanophotocatalytic degradation of organic toxic compounds: A general mechanism.*

also adsorb a larger range of solar spectra and a greater number of light quanta than several semiconducting metal oxides [15]. However, the light absorption of ZnO NPs is also limited in the ultraviolet wavelength region, similar to that of TiO2 NPs. Additionally, ZnO NPs are susceptible to photocorrosion, resulting in fast recombination of photogenerated charges and resulting in low photocatalytic efficiency [16]. Metallic doping of ZnO nanoparticles is a common method of improving their photodegradation. Researchers have evaluated a wide range of metal dopants, including anionic, cationic, rare-earth, and codopants [17]. The coupling of ZnO with other semiconductors, including Cadmium oxide (CdO) [18], Cerium dioxide ($CeO_2$) [19], Stannic dioxide ($SnO_2$) [20], $TiO_2$ [21], Graphene oxide (GO) [22], and Reduced graphene oxide (RGO) [23], has also been shown to enhance photodegradation efficiency of ZnO particles.

#### *3.2.3 Iron oxides nanoparticles*

There has been a growing interest in iron oxide nanoparticles as adsorbents for heavy metal removal due to their simplicity and availability. The nonmagnetic hematite ($Fe_2O_3$) has also been used in recent years for the removal of heavy metals. In general, nanosorbent materials can be difficult to separate and recover from contaminated water due to their small size. It is possible to separate and recover both magnetic magnetite ($Fe_3O_4$) and magnetic maghemite (-$Fe_2O_4$) easily by implementing of an external magnetic field. These materials were successfully used as sorbent materials to remove various heavy metals from water systems. Several ligands like ethylenediamine tetraacetic acid (EDTA), L-glutathione (GSH), mercaptobutyric acid (MBA), $\alpha$-thio-$\omega$-(propionic acid) hepta(ethylene glycol) (PEG-SH), and meso-2,3-dimercaptosuccinic acid (DMSA) or polymers like copolymers of acrylic acid and crotonic acid have been used to enhance the adsorption efficiency of iron oxide nanoparticles by mixing them with various metal ions [24–26]. It has been reported that a flexible ligand shell facilitates the incorporation of a wide array of functional groups into the shell as well as enabling $Fe_3O_4$ nanoparticles to retain their properties. Moreover, polymer shells have been shown to enhance the dispersibility of nanostructures and prevent particles from aggregating. Metal ions found in treated water could be "carried" by polymer molecules as binders [27].

### 3.3 Carbon nanotubes

These materials are unique in their structures and electronic properties, making them a fascinating class of materials that have attracted scientists for fundamental research and diverse applications, including sorption processes. They have a great capacity for adsorbing a wide range of contaminants, fast kinetics, large specific surfaces, and selectivity towards aromatics, making them very effective for the treatment of water and wastewater. Carbon nanotubes (CNTs) are one type of carbon nanomaterials (CNMs), but there are several others, including carbon beads, fibers, and nanoporous carbon. Among these, CNTs have gained the most attention and have progressed rapidly in recent years. Carbon nanotubes are graphene sheets rolled up in cylinders with a diameter as small as 1 nm. The remarkable properties of CNTs have made them a very attractive adsorbent. CNTs show extremely high specific surface areas and adsorption efficiencies for a wide variety of contaminants, due to their abundant porous structures. Many carbon nanotubes are combined with metals or other supports in order to improve their surface area, mechanical properties, optical properties, and electrical properties [28, 29].

The inhibitory mechanism of nanoparticles against different bacteria and fungi includes release of metal ions that interacts with cellular components through various pathways including reactive oxygen species (ROS) generation, pore formation in cell membranes, cell wall damage, DNA damage, and cell cycle arrest and ultimately inhibits the growth of cells. Therefore, they can be used to treat water and wastewater to gain usable besides drinkable water without any pathogens. However, Nanotechnology raises concerns in the public and activist groups regarding several fundamental aspects. In view of the much greater surface area to volume ratio of nanoparticles, researchers acknowledge the risk associated with nanomaterials may be different from the risk associated with bulk versions of the same material. This could lead to undeveloped and untested interactions with biological surfaces compared to bulk materials [30].

### 3.4 Nanocomposite

Recent years have seen a surge in the production of various nanocomposites. A huge amount of research has been carried out all over the world on the basis of numerous studies. According to the results, the adsorbent has good potential in removing nitrate from water quickly and effectively. Furthermore, due to its unique magnetic property, the adsorbent can be easily removed from the solution using a magnet. Real composite materials should be smooth, bulky, immobile materials that achieve nano reactivity by anchoring or impregnating a parent material structure with nanomaterials. Furthermore, treatment of water and wastewater needs nontoxic, long-term stable, low-cost materials. Further research is still necessary to obtain desired nanocomposites [31].

## 4. Type of nanomaterials in wastewater treatment

The obvious way of using nanotechnology in water and wastewater treatment is that this technology cannot be used itself, and, as a result, it has been found that nanomaterials had better be incorporated into industrial water treatment processes. A large amount of research has been carried out into the application of nanotechnology for wastewater treatment. Nanotechnology can be classified into three main groups based on the kinds of materials they use: Nano-adsorbents, Nano-catalysts, and Nano-membranes [6].

### 4.1 Nano-adsorbents

Adsorbent nanoparticles are nano-sized particles made from organic or inorganic materials that have a strong affinity for adsorbing substances. This means that they are able to remove a lot of pollutants. It is possible to use these nanoparticles in the removal of different kinds of pollutants due to their important characteristics, such as their catalytic potential, small size, high reactivity, and their higher surface energy. Adsorption processes can be distinguished among metallic nanoparticles, mixed oxide nanostructures, magnetic nanoparticles, and metal oxide nanoparticles [6].

### 4.2 Nano-catalyst

In nano-catalysis, light energy interacts with metallic nanoparticles, leading to high and wide photocatalytic activities. This treatment is gaining popularity due to its high and wide photocatalytic activity. In a photocatalytic reaction, bacteria and organic matter are destroyed by hydroxyl radicals. Materials used in nano-catalysts usually contain inorganic components like semiconductors and metal oxides. A nano photocatalyst must meet certain criteria to be considered a nanocatalytic; these include: being harmless and their concentration in water and air below the maximum permissible level; making agglomerates, precipitating, and forming ordinary particles [6].

### 4.3 Nano-membrane

A nano-membrane is responsible for separating particles from wastewater. These filters are very effective in removing dyes, heavy metals, and other contaminants. Nanotubes, nanoribbons, and nanofibers are nanomaterials used as nano-membranes.

Among nano-adsorbents, nano-catalysts, and nano-membranes, nanoparticles integrated into membranes are more convenient and useful because of the fact that not only does this process include a powerful physical treatment in it, but it also has nanoparticles to improve the quality of the treatment. Therefore, the upcoming discussion is about nano-membranes.

As nanotechnology has grown rapidly, the fabrication of nanocomposite membranes has become increasingly significant and efficient. As a result, a wide range of nanoparticles (NPs) has been examined for their implications on polymeric nanocomposite membrane engineering properties; in many cases, these NPs have improved mechanical, thermal, and antifouling properties significantly [32]. **Table 1** shows some of the membranes prepared with nanotechnology in order to prepare modified membranes. NPs are widely known to increase the mechanical and thermal properties of polymeric membranes when they are dispersed uniformly into the polymeric matrix and formed strong interfacial bonds with the matrix. Membranes are broadly used in membrane bioreactor (MBR) systems for treating various types of wastewater. As compared to conventional processes, the MBR process, which combines activated sludge with membrane filtration,

| Membrane type | Material of membrane | Nanoparticle | Main result | Reference |
|---|---|---|---|---|
| Ceramic | α-$Al_2O_3$ filters | $TiO_2$ | Rejection improved | [33] |
| Ceramic | Alumina | $Fe_2O_3$ | Rejection improved | [34] |
| Polymeric | Polypropylene (PP) | $SiO_2$ | Modified membrane fouling decreased | [35] |
| Polymeric | Cellulose acetate (CA) | Nanodiamond (ND) and ND-COOH | Mechanical, thermal, and antibacterial properties improved | [36] |
| Polymeric | Polyvinyl chloride (PVC) | Ag and Ag@$SiO_2$ | Fouling improved and flux increased | [37] |
| Polymeric | PVDF | ND and PVP-ND | Fouling and flux improved | [38] |
| Polymeric | PVDF | Ag@$SiO_2$ | Fouling and flux improved | [39] |
| Polymeric | PVDF | Ag@TiO2 | Fouling resistance and antibacterial performance enhanced | [40] |

**Table 1.**
*Uses of diverse nanoparticles in several membranes and results.*

produces a much higher effluent quality, a smaller footprint, a higher organic loading rate, and less sludge. In this process, membranes are fouled by not only organic matters but also by micro-organisms and their productions. Even though the operating conditions have a great effect on membrane fouling in MBRs, membranes' characteristics plays a key role in this regard [41].

In order to boost the hydrophobicity of polymer membranes and their antifouling properties, different techniques can be used by embedding organic nanoparticles in polymer matrices, for example. In a broad sense, as mentioned in **Table 1.**, incorporating nanoparticle in membrane matrix have these three obvious effects:

1. (NOT IN ALL CASES) Triggers to smaller and more pores on the surface of the membrane that leads to more flux and better rejection.

2. By using anti-bacterial nanoparticles and modifying operating conditions the fouling of the membrane can considerably be decreased.

3. Prevent micro-organisms and their bodies' production, which are mainly composed of proteins and carbohydrates, from attaching membrane surface and pores and causes severe fouling.

4. Decreasing frequent cleaning needs for recovery.

It should be noted that a lot of studies have been done by using one nanoparticle in a membrane matrix and found that under these circumstances releasing of the nanoparticle is regarded as a downside. Therefore, researchers try to use more than one nanoparticle in order to make them bigger and to trap them in the membrane matrix with the least releasing.

## 5. Conclusions

Nanotechnology should be utilized as a supplementary technology beside other means for wastewater treatment. The best form of using nanomaterials in this process is incorporating them with or coating them on membranes and composites. When comparing nanofilters to conventional systems, nanofilters have the following key advantages: Less pressure is required to pass water across the filter that means it dramatically decreases the operational costs, they are more efficient, and they have enormous surface areas which can be easily cleaned by back-flushing compared to conventional methods. Briefly, nanotechnology is highly improved the downside of the treatment technologies; however, it is not eliminated all problems of them. Thus, there is a long way to study and comprehend the suitable ways of the combination of the diverse treatment systems in order to reach several best treatment processes whereby human beings can save the water on Earth for more years to the next generations.

## Author details

Farid Alizad Oghyanous
Faculty of Chemical Engineering, Sahand University of Technology, Tabriz, Iran

*Address all correspondence to: fa_alizad@sut.ac.ir

## References

[1] N.P. Cheremisinoff, Handbook of water and wastewater treatment technologies, Butterworth-Heinemann, 2001.

[2] G. Tchobanoglous, F.L. Burton, H.D. Stensel, Metcalf & Eddy, Inc, Wastewater Engineering: Treatment, Disposal, Reuse (2003).

[3] C.A. Dieter, Water availability and use science program: Estimated use of water in the United States in 2015, Geological Survey, 2018.

[4] S. Adham, A. Hussain, J. Minier-Matar, A. Janson, R. Sharma, Membrane applications and opportunities for water management in the oil & gas industry, Desalination, 440 (2018) 2-17.

[5] G. Gude, Sustainable desalination handbook: plant selection, design and implementation, Butterworth-Heinemann, 2018.

[6] D.L. Schodek, P. Ferreira, M.F. Ashby, Nanomaterials, nanotechnologies and design: an introduction for engineers and architects, Butterworth-Heinemann, 2009.

[7] R.S. Kalhapure, S.J. Sonawane, D.R. Sikwal, M. Jadhav, S. Rambharose, C. Mocktar, T. Govender, Solid lipid nanoparticles of clotrimazole silver complex: an efficient nano antibacterial against Staphylococcus aureus and MRSA, Colloids and Surfaces B: Biointerfaces, 136 (2015) 651-658.

[8] B. Borrego, G. Lorenzo, J.D. Mota-Morales, H. Almanza-Reyes, F. Mateos, E. López-Gil, N. de la Losa, V.A. Burmistrov, A.N. Pestryakov, A. Brun, Potential application of silver nanoparticles to control the infectivity of Rift Valley fever virus in vitro and in vivo, Nanomedicine: Nanotechnology, Biology and Medicine, 12 (2016) 1185-1192.

[9] C. Krishnaraj, R. Ramachandran, K. Mohan, P. Kalaichelvan, Optimization for rapid synthesis of silver nanoparticles and its effect on phytopathogenic fungi, Spectrochimica Acta Part A: Molecular and Biomolecular Spectroscopy, 93 (2012) 95-99.

[10] M. Rivero-Huguet, W.D. Marshall, Reduction of hexavalent chromium mediated by micron-and nano-scale zero-valent metallic particles, Journal of Environmental Monitoring, 11 (2009) 1072-1079.

[11] K. Guesh, A. Mayoral, C. Marquez-Alvarez, Y. Chebude, I. Diaz, Enhanced photocatalytic activity of TiO2 supported on zeolites tested in real wastewaters from the textile industry of Ethiopia, Microporous and Mesoporous Materials, 225 (2016) 88-97.

[12] A. Yamakata, J.J.M. Vequizo, Curious behaviors of photogenerated electrons and holes at the defects on anatase, rutile, and brookite TiO2 powders: A review, Journal of Photochemistry and Photobiology C: Photochemistry Reviews, 40 (2019) 234-243.

[13] K. Umar, A.A. Dar, M. Haque, N.A. Mir, M. Muneer, Photocatalysed decolourization of two textile dye derivatives, Martius Yellow and Acid Blue 129, in UV-irradiated aqueous suspensions of Titania, Desalination and Water Treatment, 46 (2012) 205-214.

[14] N. Daneshvar, D. Salari, A. Khataee, Photocatalytic degradation of azo dye acid red 14 in water on ZnO as an alternative catalyst to TiO2, Journal of photochemistry and photobiology A: chemistry, 162 (2004) 317-322.

[15] A. Janotti, C.G. Van de Walle, Fundamentals of zinc oxide as a semiconductor, Reports on progress in physics, 72 (2009) 126501.

[16] C. Gomez-Solís, J. Ballesteros, L. Torres-Martínez, I. Juárez-Ramírez, L.D. Torres, M.E. Zarazua-Morin, S.W. Lee, Rapid synthesis of ZnO nano-corncobs from Nital solution and its application in the photodegradation of methyl orange, Journal of Photochemistry and Photobiology A: Chemistry, 298 (2015) 49-54.

[17] K.M. Lee, C.W. Lai, K.S. Ngai, J.C. Juan, Recent developments of zinc oxide based photocatalyst in water treatment technology: a review, Water research, 88 (2016) 428-448.

[18] M. Samadi, A. Pourjavadi, A. Moshfegh, Role of CdO addition on the growth and photocatalytic activity of electrospun ZnO nanofibers: UV vs. visible light, Applied surface science, 298 (2014) 147-154.

[19] I.-T. Liu, M.-H. Hon, L.G. Teoh, The preparation, characterization and photocatalytic activity of radical-shaped CeO2/ZnO microstructures, Ceramics International, 40 (2014) 4019-4024.

[20] M.T. Uddin, Y. Nicolas, C. Olivier, T. Toupance, L. Servant, M.M. Muller, H.-J. Kleebe, J. Ziegler, W. Jaegermann, Nanostructured $SnO_2$–ZnO heterojunction photocatalysts showing enhanced photocatalytic activity for the degradation of organic dyes, Inorganic chemistry, 51 (2012) 7764-7773.

[21] H.R. Pant, C.H. Park, B. Pant, L.D. Tijing, H.Y. Kim, C.S. Kim, Synthesis, characterization, and photocatalytic properties of ZnO nano-flower containing TiO2 NPs, Ceramics International, 38 (2012) 2943-2950.

[22] K. Dai, L. Lu, C. Liang, J. Dai, G. Zhu, Z. Liu, Q. Liu, Y. Zhang, Graphene oxide modified ZnO nanorods hybrid with high reusable photocatalytic activity under UV-LED irradiation, Materials Chemistry and Physics, 143 (2014) 1410-1416.

[23] X. Zhou, T. Shi, H. Zhou, Hydrothermal preparation of ZnO-reduced graphene oxide hybrid with high performance in photocatalytic degradation, Applied surface science, 258 (2012) 6204-6211.

[24] Y. Lei, F. Chen, Y. Luo, L. Zhang, Three-dimensional magnetic graphene oxide foam/Fe 3 O 4 nanocomposite as an efficient absorbent for Cr (VI) removal, Journal of Materials Science, 49 (2014) 4236-4245.

[25] L. Tan, J. Xu, X. Xue, Z. Lou, J. Zhu, S.A. Baig, X. Xu, Multifunctional nanocomposite Fe 3 O 4@ $SiO_2$–mPD/SP for selective removal of Pb (ii) and Cr (vi) from aqueous solutions, RSC advances, 4 (2014) 45920-45929.

[26] A.-F. Ngomsik, A. Bee, D. Talbot, G. Cote, Magnetic solid–liquid extraction of Eu (III), La (III), Ni (II) and Co (II) with maghemite nanoparticles, Separation and purification technology, 86 (2012) 1-8.

[27] R.A. Khaydarov, R.R. Khaydarov, O. Gapurova, Water purification from metal ions using carbon nanoparticle-conjugated polymer nanocomposites, Water research, 44 (2010) 1927-1933.

[28] A. Chatterjee, B. Deopura, Carbon nanotubes and nanofibre: an overview, Fibers and Polymers, 3 (2002) 134-139.

[29] M.M. Khin, A.S. Nair, V.J. Babu, R. Murugan, S. Ramakrishna, A review on nanomaterials for environmental remediation, Energy and Environmental Science, 5 (2012) 8075-8109.

[30] J. Singh, K. Vishwakarma, N. Ramawat, P. Rai, V.K. Singh, R.K. Mishra, V. Kumar, D.K. Tripathi, S. Sharma, Nanomaterials and microbes' interactions: a contemporary overview, 3 Biotech, 9 (2019) 1-14.

[31] M. Peyravi, M. Jahanshahi, A. Rahimpour, A. Javadi, S. Hajavi, Novel

thin film nanocomposite membranes incorporated with functionalized TiO2 nanoparticles for organic solvent nanofiltration, Chemical Engineering Journal, 241 (2014) 155-166.

[32] J. Dasgupta, S. Chakraborty, J. Sikder, R. Kumar, D. Pal, S. Curcio, E. Drioli, The effects of thermally stable titanium silicon oxide nanoparticles on structure and performance of cellulose acetate ultrafiltration membranes, Separation and Purification Technology, 133 (2014) 55-68.

[33] B. Zhu, Y. Hu, S. Kennedy, N. Milne, G. Morris, W. Jin, S. Gray, M. Duke, Dual function filtration and catalytic breakdown of organic pollutants in wastewater using ozonation with titania and alumina membranes, Journal of membrane science, 378 (2011) 61-72.

[34] L. De Angelis, M.M.F. de Cortalezzi, Improved membrane flux recovery by Fenton-type reactions, Journal of Membrane Science, 500 (2016) 255-264.

[35] M. Ahsani, R. Yegani, Study on the fouling behavior of silica nanocomposite modified polypropylene membrane in purification of collagen protein, Chemical engineering research and design, 102 (2015) 261-273.

[36] H. Etemadi, R. Yegani, V. Babaeipour, Study on the reinforcing effect of nanodiamond particles on the mechanical, thermal and antibacterial properties of cellulose acetate membranes, Diamond and Related Materials, 69 (2016) 166-176.

[37] A. Behboudi, Y. Jafarzadeh, R. Yegani, Enhancement of antifouling and antibacterial properties of PVC hollow fiber ultrafiltration membranes using pristine and modified silver nanoparticles, Journal of Environmental Chemical Engineering, 6 (2018) 1764-1773.

[38] M. Javadi, Y. Jafarzadeh, R. Yegani, S. Kazemi, PVDF membranes embedded with PVP functionalized nanodiamond for pharmaceutical wastewater treatment, Chemical Engineering Research and Design, 140 (2018) 241-250.

[39] M. Ahsani, H. Hazrati, M. Javadi, M. Ulbricht, R. Yegani, Preparation of antibiofouling nanocomposite PVDF/Ag-SiO2 membrane and long-term performance evaluation in the MBR system fed by real pharmaceutical wastewater, Separation and Purification Technology, 249 (2020) 116938.

[40] J.R. Mishra, S.K. Samal, S. Mohanty, S.K. Nayak, Polyvinylidene fluoride (PVDF)/Ag@ TiO2 Nanocomposite Membrane with Enhanced Fouling Resistance and Antibacterial Performance, Materials Chemistry and Physics (2021) 124723.

[41] F.A. Oghyanous, H. Etemadi, R. Yegani, The effect of sludge retention time and organic loading rate on performance and membrane fouling in membrane bioreactor, Journal of Chemical Technology and Biotechnology, 96 (2021) 743-754.